This book is the first to provide a comprehensive coverage of the field of liquid crystalline polymers. It is a graduate text, intended for those who are studying liquid crystalline polymers within the disciplines of chemistry, physics or materials science. It will also serve as a standard reference for all involved, at whatever level, with these materials, providing the necessary framework within which to approach the growing literature on the subject.

Both main chain and side chain polymer variants of the thermotropic and lyotropic types of liquid crystalline polymer are considered. After an introduction to the terminology of the subject, the book examines both the methods of synthesis and the properties of these materials, including discussion on packing, the nature of defects, rheological properties and the response to external fields. A review of theoretical approaches is also included, and the book finally examines the potential applications of liquid crystalline polymers, notably for electro-optical devices and as high strength materials.

The book will be of importance to all students and researchers in any discipline who are studying liquid crystalline polymers, and will be accessible both to those approaching the subject from a background of conventional polymer science and to those with experience of small molecule liquid crystalline polymers.

Liquid crystalline polymers

Cambridge Solid State Science Series

EDITORS:

Professor R. W. Cahn, FRS
University of Cambridge

Professor E. A. Davis
Department of Physics, University of Leicester

Professor I. M. Ward, FRS
Department of Physics, University of Leeds

A. M. DONALD

Cavendish Laboratory, University of Cambridge

A. H. WINDLE

Department of Materials Science and Metallurgy, University of Cambridge

Liquid crystalline polymers

CAMBRIDGE
UNIVERSITY PRESS

0526 7031

CHEMISTRY

Published by the Press Syndicate of the University of Cambridge
The Pitt Building, Trumpington Street, Cambridge CB2 1RP
40 West 20th Street, New York, NY 10011–4211, USA
10 Stamford Road, Oakleigh, Victoria 3166, Australia

© Cambridge University Press 1992

First published 1992

Printed in Great Britain at the University Press, Cambridge

British Library cataloguing in publication data
Donald, A. M.
Liquid crystalline polymers.
1. Polymers. Liquid crystals
I. Title II. Windle A. H.
547.7

Library of Congress cataloguing in publication data
Donald, A. M. (Athene Margaret), 1953–
Liquid crystalline polymers / A. M. Donald and A. H. Windle.
    p.   cm. – (Cambridge solid state science series)
Includes bibliographical references (p. 000) and index.
ISBN 0 521 30666 3 (hard)
1. Polymer liquid crystals. I. Windle, A. H. II. Title.
III. Series.
QD923.D66   1992
530.4′29–dc20 90–2567 CIP

ISBN 0 521 30666 3 hardback

Contents

QD 923
D66
1992
CHEM

Preface

Liquid crystalline polymers lead one into the heartland of interdisciplinary science, where it is an art in itself to thrive without becoming a Jack of all trades and master of none. To explore new frontiers requires depth, but in this subject probably more than any other there is need for a breadth of background, and often a breadth not catered for within the traditional divisions of school and college science education. The subject lies where organic chemistry, physical chemistry, physics and materials science meet. A physicist will discover the need to appreciate something of the chemical diversity of the molecular world, and will begin to become familiar with the chemical language of the trade. A chemist, skilled in the synthesis of new liquid crystalline molecules, will become increasingly familiar with the need of materials people for samples in kilogram rather than milligram quantities, and with the demands of the physicist that the samples be ever purer and better characterized. Indeed, any scientist meeting the field for the first time will almost certainly be confronted with some unfamiliar areas which must be assimilated and understood in depth. This book is written as a handbook for anyone entering the liquid crystalline polymer arena, from whatever background. We hope it will serve the graduate student embarking on any project associated with these novel materials, the academic desiring a broader knowledge and the scientist whose research efforts become focused into the field. The authors also intend it as a reference text for any course at the undergraduate or masters degree level which includes liquid crystalline polymers within its syllabus. Recognizing the likely diversity of experience of its readers, the authors have taken a liberty with the second chapter, *Terminology and concepts*. This they present encyclopaedically, with a wide range of topics appearing in alphabetical order. It is unlikely to be read before the remainder of the book, and is not really meant to be. It is offered as a resource to serve areas in which a reader's background knowledge may need support, and is there to be dipped into as needed.

Liquid crystalline polymers, as a subject area, has arisen from the interaction of two well-established sciences. The more mature of these is liquid crystalline science, now a little over 100 years old (see Chapter 1) which comparatively recently experienced a great upsurge of interest in its own right as the application of liquid crystals to display devices fuelled it with funds and new challenges. Polymer science, by comparison, is somewhat younger, albeit the fore-runner of the plastics industry. The recognition that there are polymers which show liquid crystallinity has, for the device scientist, opened up the possibility of attaching mesogenic molecules onto or into polymer chains, and led to a wide range of new electro-optical opportunities. On the other hand, the polymer field now has available to it a range of new materials with novel processing capabilities and with the prospect of unique (and marketable!) combinations of properties. Indeed in the heady days of the mid 1980s, there was hardly a large chemical company which was not active in the field, although hopes of a rapid return on investment meant that the subject tended to be oversold. Six years and a world recession later, there are perhaps only four major chemical companies sharing the market for structural liquid crystalline polymers, while device applications are in the advanced development stage. The basic research effort worldwide remains at a high level and is maturing. Integrated teams now follow programmes through from the synthesis of new molecules, to chemical, physical and microstructural characterization, then to properties and finally to applications, with feedback loops established to the chemistry. This is the current climate in which this book appears. It is perhaps surprising, but, as far as the authors are aware, it is the first dedicated text book to be published on the subject. There are, of course, a number of multi-author, edited volumes of papers and articles which continue to serve the field in providing up to date research reviews. Indeed, the latest of these, edited by Ciferri and published by VCH (1991) is too recent to be referenced formally in this text but is warmly commended to the reader.

The *contents* pages have already mapped this book. After a brief history of liquid crystalline polymers and the background chapter on terminology and concepts already described, Chapter 3 concentrates on relating the stability of liquid crystalline polymers, as a function of temperature and solvent content, to their molecular architecture. It builds strongly on the established knowledge of small molecule liquid crystalline phases, and will be an especially appropriate entry point to any who are familiar with that side of the subject. While Chapter 3 focuses on the influence of chemical detail on stability, Chapter 4 takes the reader through the different theories which describe liquid crystalline polymers using, deliberately, the simplest of molecular models. Attention is given to the search for the critical molecular parameters which are the essence of chain mesogenicity, and which determine transition temperatures and microstructures both in thermotropic melts and lyotropic solutions. Chapter 5 examines aspects of local molecular order in more detail and introduces the reader to the Friedelian classification of polymeric mesophases. Chapter 6 moves up in scale to the continuum level where the microstructure of liquid crystalline polymer phases are accounted for in terms of distortion energies and defects within the elastic distortion fields, while Chapter 7 represents a further increase in scale, describing the response to external applied fields. While most emphasis is placed on flow fields and the associated rheological issues, there are also five pages or so devoted to the effect of electromagnetic fields. In writing Chapter 8 on practical aspects of liquid crystalline polymers, the authors have sought to provide a snapshot of current applications of the materials. They have done this in the knowledge that this chapter is likely to age rather more quickly than the preceding ones, but it is a risk taken in the belief that an appreciation of the applications of the science to technology and thus to commerce, makes the best possible context for fundamental research in providing a spur to new ideas.

In addition to the references, by chapter at the end of the book, there are also indices both of symbols used and of chemical formulae of the vast majority of molecules referred to in the book. The molecules are indexed by numbers in angle brackets, the references in square brackets, while the equations are numbered in round brackets. The units used are SI, although some more esoteric ones such as N tex^{-1}, are briefly introduced, more to prepare the reader for what may be encountered elsewhere than to serve the needs of the text itself. In addition e.s.u. still tends to be used in electromagnetic experiments. Particularly, we ask the reader's bilingual indulgence in the matter of temperature units. While we have used Kelvin (K) where there are no other constraints, so many data are still published in centigrade units (°C) that we have not converted reported values, either in figures or in the text.

With the book now complete it is our great pleasure to thank those who have helped us to make it appear at all. In the first place, it was Professor Ian Ward who recognized the need for such a text and encouraged us to go ahead. We are especially grateful to him for his leadership, support and constructive comments on the manuscript, and his unfailing patience. We must also acknowledge with gratitude, all those who have influenced us in our scientific careers and thus indirectly contributed here. In particular we wish to thank our various mentors in (respectively) Materials Science at Cornell University and the Department of Physics at Bristol University, as well as here at Cambridge. We also recognize the contribution of our University through its enlightened attitude to the sabbatical principle. Our research students, past and present, have given much, not only in their tolerance of our reduced availability at times, but also through their candid critique of various drafts of chapters. Our respective spouses, Matthew and Janet, and our families, have a big part in this too. For their love and assistance over six summers, we thank them.

Athene Donald
Alan Windle

1   Liquid crystalline polymers: a brief history

In March 1888, a young botanist called Friedrich Reinitzer, wrote to Otto Lehmann who was professor of physics at Aachen. He described observations, which he published that year [1.1], that esters of cholesterol appeared to have two melting points between which the liquid showed iridescent colours and birefringence. It was sensible to consult Lehmann who had worked for some time in the field of crystal transitions, and above all, had developed a polarizing microscope with a hot stage which was to become a central feature of much of his research. He confirmed Reinitzer's observations, and postal collaboration between the two over the next few years laid the foundations of liquid crystal science. By 1889 Lehmann was describing the material as 'flowing crystals' and the following year as 'crystalline liquids' [1.2]. From this point onwards, he was making most of the running and gradually became the hub of a wider collaboration. He spread interest beyond Germany, notably to France and Holland. Liquid crystal science continued to mature, and synthetic chemists, particularly Vorlander at Halle, demonstrated the principles of molecular design which underpin the field. There was considerable debate between George Friedel working in Strasbourg, and Lehmann and Vorlander, as to the existence of different types of liquid crystalline phase. However, the issue eventually reduced to one of semantics, as it was recognized that liquid crystallinity does indeed appear in several different guises.

1922, the year of Lehmann's death, marked two important developments. Friedel published a monumental paper [1.3] which cleared up much of the terminology of the subject. Drawing on Greek roots he introduced the terms, 'nematic', 'smectic', and also 'mesophase', which he suggested should supplant 'liquid crystals' as the generic title. The second development was within a different scientific community altogether. Staudinger was mounting his challenge to the chemistry establishment with the notion that long-chain molecules could actually exist. Vorlander seemed rapidly at ease with the idea of polymer molecules, and as early as 1923 he began to pursue the concept of a liquid crystalline polymer, systematically studying the effect of increasing the length of liquid crystalline molecules [1.4]. He synthesized rods with one, two and three benzene rings *para* linked through ester groups, and noted their increasing transition temperatures. His discussion of polymers based on these units is illustrated with a note of the synthesis of poly(*p*-benzamide). He found that the polymer would not melt, reporting it as a birefringent powder which chars without softening. In many respects, Vorlander can justifiably be seen as the father of polymeric liquid crystals.

While the problem of synthesizing polymers which were mouldable in the liquid crystalline phase was only solved finally by Jackson some 50 years later, there was increasing interest in rod-like macromolecules which occurred in nature and formed liquid crystalline phases when in solution (lyotropic phases). A key observation was that of Bawden and Pirie in 1937 [1.5] who noted that solutions of tobacco mosaic virus (TMV), a stiff rod-like molecule a few thousand ångströms long, separated into two liquid phases one of which was birefringent. They also showed that the birefringent phase had a higher concentration of polymer. This line of approach continued slowly over the next 20 years or so, other rod-like viruses such as the cucumber virus and the potato virus being added to the list. Bernal and Fanchuchen (1941) [1.6] examined the liquid crystalline phase with X-ray diffraction while Oster (1950) [1.7] carried out light scattering measurements.

Activity now moved to the laboratories of Courtaulds at Maidenhead in England, where as a part of a programme to develop synthetic silk they produced a series of polyglutamates, particularly poly(γ-benzyl-L-glutamate) (PBLG). These polymers showed liquid crystalline phases in solution and formed the basis of a particularly detailed study of structure and phase equilibria published by Robinson in 1956 [1.8]. Although PBLG fibres were spun from the lyotropic phase by Ballard (1958) [1.9] they were not developed into a product. The first successful commercialization of liquid crystalline polymers was achieved by Du Pont with their aromatic amide fibre, marketed as Kevlar. This fibre which has outstanding tensile properties, was developed by Kwolek and her team through the 1960s [1.10]. It was spun from a lyotropic solution using comparatively fierce solvents, such as 100% sulphuric acid, which were then washed out from the fibre. While the liquid crystalline nature of the aromatic amide solutions was recognized by those working with them, it is interesting that developments of semi-rigid chain polymers in other areas, such as those involving modified celluloses, proceeded without the realization that some of their particularly interesting properties were the result of liquid crystallinity.

Vorlander's effort to make a liquid crystalline polymer which would be stable without the addition of solvents was frustrated by the fact that the crystal melting point of rigid chain crystals increases rapidly with increasing chain length. Research by Jackson and co-workers in the Kingsport, Tennessee, laboratories of Eastman Kodak in the 1970s led to the synthesis of rigid, random copolymers of aromatic polyesters. In this way the melting point was much reduced, if not eliminated altogether, and the polymer could be processed as a liquid crystalline melt without risk of degradation. The contribution of Jackson has been seminal to the field. In addition to patents he has published a series of papers, the first of which, with Kuhfuss in 1976 [1.11], really brought liquid crystalline polymers to the attention of the polymer community at large. Eastman Kodak were also very generous with research samples and within a few years interest in the field had blossomed in industrial and academic circles worldwide. There are now, in the 1990s, a number of mouldable liquid crystalline materials marketed by various chemical companies, the Vectra range from Hoechst Celanese being a leading example.

The gradual development of liquid crystalline polymers, from the early work on TMV in solution to their exploitation as bulk plastic and fibre materials, has been underpinned by advances in theories which explain the detailed observations. Onsager (1949) [1.12] was the first to introduce a theory of liquid crystalline polymers. By using a virial expansion, he explained the observed biphasic nature of solutions of rigid molecules. A somewhat different approach was that of Flory (1956) [1.13] who adapted his lattice model of conventional polymers to the rigid-rod situation. He was able to predict the form of the polymer/solvent phase diagram determined experimentally by Robinson, and his school developed the work over a number of years to the point where it became possible to treat semi-rigid worm-like molecules and also account for specific interchain interactions.

Thus far, the development of liquid crystalline polymers has been traced independently from progress in the field of conventional (small-molecule) liquid crystals. The development of liquid crystal displays in the 1960s, produced a veritable explosion of research activity which often involved chemists and physicists working closely together. New insights have been gained and better molecules developed. The involvement of polymeric materials has been largely a matter of adding active rod-like groups, similar to conventional liquid crystalline molecules, onto flexible polymer chains. They are added either as side chains or into the backbone itself. The key advance in the development of side-chain liquid crystalline polymers was the realization by Ringsdorf, Finkelmann and coworkers [1.14] that rod-like side chains would only readily form mesophases if they were decoupled from the backbone to which they were attached by means of a short length of flexible chain acting as a spacer. The activity, and in particular, the field orientability of the rigid groups is modified because of their attachment to the chain. Time constants can be lengthened which may be of possible advantage to storage displays. However, the main benefits accrue from the fact that the backbone renders the material solid and formable into films or fibres without the need for containment in a cell. Furthermore, it means that the material can often be quenched to form a glass, which avoids the strong optical scattering associated with crystallization of conventional liquid crystals. The development of side-chain materials continues apace, and is now marked by a recent collection of papers edited by McArdle [1.15].

It is becoming clear that 'self-assembly', where molecules are designed so that they organize themselves into larger scale structures in order to achieve special properties, will be a significant objective in materials science in the next century. The fact that liquid crystallinity itself is a form of orientational self assembly, coupled with the fact that the molecules in a mesophase can be steered by external fields, means that the principles underlying the science of liquid crystalline polymers can only increase in significance in the years ahead.

2 Terminology and concepts (in alphabetical order)

While both polymer science and liquid crystalline science are multidisciplinary fields in their own right, their combination in the study of liquid crystalline polymers creates a subject area which requires an unusually wide range of background expertise. In preparing this book, we recognize that the subject is demanding in that its full appreciation will almost certainly involve the reader in some unfamiliar terminology and new concepts. This chapter is included in order to lighten that task. It provides a number of background topics arranged alphabetically which can be referred to as required. Almost inevitably, some of the sections will appear rather banal where they correspond to areas in which the reader is already knowledgeable, however, that is perhaps a small price to ensure that the text is accessible to all who have taken some branch of physical science at university level. This type of arrangement also relieves the main chapters of excessive background material, and, where possible, key terms used in the text are explained more fully here.

Bend (see **Distortions**).

Chain dimensions The starting point for much of the physics of polymer molecules is the simple, freely jointed, random chain. In it, the links are all of identical length, l_0, and there is absolutely no restriction on the relative orientations of adjacent links, even to the extent of permitting the possibility of one link lying back on top of its predecessor. Such a model chain obeys simple physical principles as it is, in effect, a random walk in three dimensions. The total length of the chain, measured along its trajectory is known as the *contour length, L*, and equals $n l_0$ where n is the number of links; this length would also be the direct distance between the ends in the event of the chain being pulled out straight. In the absence of external influences, the chain is free to adopt many different *conformations* which will correspond to different paths and hence different end-to-end distances. Thus the direct distance between the ends of a random chain must be expressed as an average over many different conformations. The mean square of the end-to-end distance, $\langle r^2 \rangle$, is given by:

$$\langle r^2 \rangle = n l_0^2$$

or $\qquad \langle r^2 \rangle^{\frac{1}{2}} = n^{\frac{1}{2}} l_0$

Another measure of the dimensions of a random chain is the *radius of gyration*, R_g. It is defined as the second moment of the distribution of the chain units about the centre of mass of the molecule. Averaged over many conformations we have the root mean square radius of gyration:

$$\langle R_g^2 \rangle^{\frac{1}{2}} = \frac{1}{n} \left(\sum_{i=1}^{n} \langle t_i^2 \rangle \right)^{\frac{1}{2}}$$

where t_i is the distance from the centre of gravity of the chain to the ith link, and $\langle \ \rangle$ indicates the average over many different conformations of the chain. In the limiting case of very high molecular weight (denoted by subscript ∞) R_g can be related simply to the mean square end-to-end distance by:

$$\langle R_g^2 \rangle_\infty = \tfrac{1}{6} \langle r^2 \rangle_\infty$$

Consideration of the dimensions of completely flexible chains may seem far removed from an understanding of rigid liquid crystalline polymer molecules. However, the degree of flexibility which a chain does have is often critical in determining whether it will form a mesophase or not. Semi-rigid chains are often described as *worm-like chains*. But even chains like these can be thought of in terms of a freely jointed chain. It is merely a matter of choosing the length of the imaginary links, l_k, so that the mean square end-to-end length of the semi-rigid worm satisfies the relation:

$$\langle r^2 \rangle = n_k l_k^2$$

Such a freely jointed equivalent chain is known as a *Kuhn chain*, and l_k as the Kuhn link length.

A useful measure of the straightness of any chain molecule is a parameter known as the *persistence length*, which is designated q in this book. It is defined as the average projection of the end-to-end distance of an infinite chain in the direction of the first link. If the chain is freely jointed and one sets the origin for the measurement of the persistence length at the start of the first link, then $q = l_0$, or in the case of a Kuhn chain, $q = l_k$. However, it is apparent that a persistence length measurement on a worm-like chain would have no special starting point, although the persistence length of the equivalent Kuhn chain will depend on the origin chosen, and needs to be expressed as the mean value over many starting points chosen at random. As the origin is moved along the first link, the value of q decreases from l_k to zero, hence its mean value is $l_k/2$. The relation between the mean square end-to-end length and the persistence length can be shown to be:

$$\langle r^2 \rangle = 2q^2[x - 1 + \exp(-x)]$$

where $x = L/q$; (L is the chain contour length, equal to $n_k l_k$).

Where the total chain length (L) is considerably greater than the persistence length (q), the relation reduces to $\langle r^2 \rangle = 2q^2 L/q = 2qn_k l_k$, and hence as $\langle r^2 \rangle = n_k l_k^2$, $q = l_k/2$ as above.

Many general textbooks provide accounts of chain physics in greater detail, and the reader is referred to Elias [2.1] amongst others.

Chiral (see **Friedelian classes**).

Cholesteric (see **Friedelian classes**).

Common tangent construction (see **Solution thermodynamics**).

Configuration and conformation 'Configuration' and 'conformation' are aspects of polymer chain structure. Their usage is not homogeneous throughout the polymer literature and there is ground for some confusion. However, for the purpose of this text they are defined as follows. *Configuration* describes the way the chain is put together at the polymerization stage. It is dependent on the pattern in which *stereoisomers* of the repeat unit are joined together, whether the units are joined head–head and tail–tail or head–tail, and whether different types of units are mixed to make a *copolymer*. In fact, the configuration of a molecule is built in at polymerization, and cannot be changed without breaking covalent bonds. No amount of rotation about bonds can affect it.

On the other hand, the *conformation* of a molecule is the arrangement of a molecule of given configuration, when there are particular settings of all the rotation angles about bonds and of bond lengths or dihedral angles between bonds which may be distorted. The *conformation* of a molecule changes as a consequence of its environment and its temperature, and in response to externally applied fields.

Conformation (see **Configuration and conformation**).

Contour length (see **Chain dimensions**).

Copolymers Where two or more different monomer units share the same type of chemical functionality, they can be polymerized together within each molecule. Taking the simplest example of polymers containing two types of units, it is clear that the molecules can vary in their average composition as well as in the distribution of the sequences. A copolymer consisting of two types of unit, A and B, in equal proportions, can be organized as either:

an *alternating copolymer,*

ABABABABABABABABABABABABABABAB

a *random copolymer,*

AABBABABBBAAAABABABAAAAABBBBABAB

a *block copolymer,*

AAAAAAAAAABBBBBBBBAAAAAAABBBBBBBBB

or a *graft copolymer,*

AAAAAAAAAAAAAAAAAAAAAAAAAAAAAAAAAA
B   B   B
B   B   B
B   B   B
B   B   B
B   B   B
B   B   B

In the case of the random copolymer there will be a statistical distribution of homopolymer sequence lengths some of which will be quite long.

Random copolymerization is used extensively in the design of aromatic main-chain thermotropic polymers where it is particularly effective in reducing the crystal melting point down to a manageable range.

Crankshaft rotation The conformational freedom of an isolated chain molecule depends on the extent to which backbone bonds are free to rotate. This freedom can depend on the intrinsic rotation potential around the bond, and steric hindrance to rotation due to non-bonded interactions, and of course the temperature. When the chain forms part of a condensed phase however, chain mobility will in general require the cooperation of other chains in the immediate neighbourhood. At one extreme one can envisage small motions, perhaps the rotation of a methyl side group around its attachment bond, which will require little if any cooperative movements from adjacent molecules, and thus will occur almost as easily in a condensed phase, as in an isolated molecule. On the other hand, rotation about a backbone bond which is not parallel to the local chain axis must necessarily cause the chain to sweep out an arc in space, so it is to be expected that such motions will be highly inhibited in the condensed state. *Crankshaft* motions of the backbone enable substantial portions of the chain to move, while minimizing the swept volume and thus hindrance from the surrounding molecules. A crankshaft rotation is the rotation of a portion of a backbone between two bonds which are approximately collinear with each other. Figure 2.1(*a*) shows a general example, while Fig. 2.1(*b*) shows a portion of a thermotropic random copolyester ⟨32⟩, and illustrates how a substantial number of aromatic groups can be involved in the motion. In fact, the rotation of one *para*-linked phenylene group about its attachment bonds can be considered as yet another example of crankshaft motion (Fig. 2.1(*c*)).

Degree of polymerization (DP) (see **Molecular weight**).

Differential scanning calorimetry (DSC) Differential scanning calorimetry is the standard technique used for measuring transition temperatures and also enthalpy changes at these transitions. The principles are simple, although to obtain quantitative data for the enthalpy change, ΔH, may not always be straightforward in practice. The basic equipment consists of two pans, one a reference pan, the

Fig. 2.1 Illustration of crankshaft type motions giving backbone mobility. (*a*) Rotations about collinear bonds (small arrows) in a simple —(CH₂)— chain in a helical conformation. The large arrow indicates the motion associated with the rotation. (*b*) Possible crankshaft process in a sequence from a thermotropic random copolyester such as ⟨32⟩. (*c*) Rotation of a para-linked phenylene group about its attachment bonds.

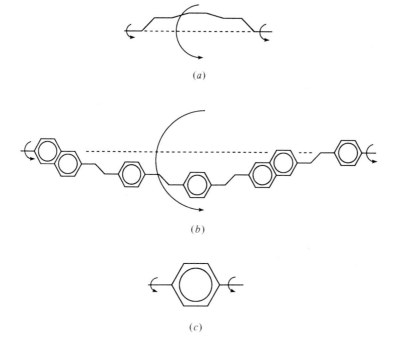

(*a*)

(*b*)

(*c*)

other containing a few milligrams of the material under study. Heat is supplied to the two pans so that their temperatures are maintained equal as the temperature is raised. (The technique can also be used for cooling cycles.) The differential power input thus indicates the thermal requirements of the specimen, and the output of the equipment is a plot of (dH/dt) against time, and thus in the case of constant heating or cooling rates, effectively dH/dT, against temperature.

Phase transitions, such as liquid crystal → isotropic, or crystal → nematic, which require the input of additional thermal energy show up as endothermic peaks on heating. A number of such peaks are apparent in the DSC trace of the thermotropic copolyester ⟨33⟩ shown in Fig. 2.2. The area under each peak (allowing for suitable baseline corrections) yields a value of the enthalpy change at the transition. The position of the maximum of the peak is commonly taken as *the* transition temperature, although this can shift with heating rate. Commonly, the peak is shifted downwards on cooling, relative to its position on heating, on account of both a machine hysteresis, which can be calibrated out, and supercooling (or possible superheating) of the transition itself. A peak which is an endotherm on heating will of

Fig. 2.2 Differential scanning calorimetry (DCS) trace of the liquid crystalline random copolyester ⟨33⟩. The heating rate was 20 °C/min. The 'step' at 70 °C corresponds to the glass transition temperature, the endotherms at 190 °C, 250 °C and 340 °C correspond to transitions in different crystalline components. The large endotherm at 420 °C can be identified with the mesophase → isotropic transition although the sample is beginning to degrade at this temperature, and any detailed interpretation must be made with caution. (Courtesy, Prof. C. Viney.)

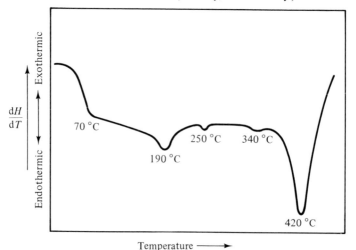

course be an exotherm on cooling and vice versa. The *glass transition* temperature, T_g, can also be measured and appears as a step on the DSC trace which corresponds to a change in slope of the enthalpy–temperature relationship. An example of a glass transition is shown on Fig. 2.2. It is the step at 70 °C. Because of overshoot effects at T_g, analogous to those seen in plots of specific volume against temperature (Fig. 2.10), there is sometimes an endothermic peak at the top of the T_g step, especially when high heating rates are used. It is important that overshoot peaks are not confused with endotherms which represent thermodynamic phase transitions.

Diffraction Diffraction is one of the most powerful techniques for revealing the internal structure of materials. Where the radiation wavelength is of the order of the intermolecular spacing, diffraction patterns can give readily interpretable information about the relative molecular positions, the presence of long range positional order, the quality of preferred orientation with respect to an external axis, and details of crystal structure and perfection. Suitable radiations are X-ray, electron and neutron. While X-rays are probably the most convenient and widely used, requiring comparatively unsophisticated equipment and sample dimensions on the millimetre scale, electrons and neutrons have advantages in particular situations. Electrons, because they are charged particles, interact much more strongly with matter and transmission diffraction experiments are only successful on very thin sections, of the order of 100 nm thick for organic materials. However, diffraction is an integral part of electron microscopy, and it is possible to directly correlate diffraction patterns with images, greatly enhancing the usefulness of both methods. Neutron diffraction is especially useful where it is desirable to examine the form of a particular molecule in an environment of similar ones. Neutrons, because they are scattered by atomic nuclei rather than electrons, are able to show contrast between different isotopes. The incorporation of a small proportion of molecules in which the hydrogen atoms have been replaced by deuterium, means that these molecules will 'stand out' from their neighbours as far as neutron diffraction is concerned, while being all but indistinguishable in chemical terms. It is thus possible, for example, to determine the trajectory of a molecule within a solid or liquid phase.

One of the central relations of crystal diffraction is the well-known Bragg equation.

$$n\lambda = 2d \sin \theta_B$$

where n is the 'order' of diffraction, λ the radiation wavelength, d the perpendicular spacing of the diffracting planes and θ_B the diffraction angle, as shown in Fig. 2.3(*a*). The equation underlines the fact that the beam is diffracted through a defined angle by interacting with a set of standing waves in electron density, i.e. the planes. The geometric restrictions of the Bragg relationship mean that a monochromatic beam directed at a perfect single crystal in an arbitrary orientation is unlikely to produce any diffracted beams at all. However, if the sample consists of many crystals in all possible orientations, a sufficient proportion will be correctly oriented to diffract. The resultant transmission diffraction pattern appears as a set of concentric circles,

Fig. 2.3 (*a*) Definition of terms, θ_B, λ and d, used in the Bragg equation. (*b*) Illustration of diffraction from a sample consisting of many crystallites in random orientation.

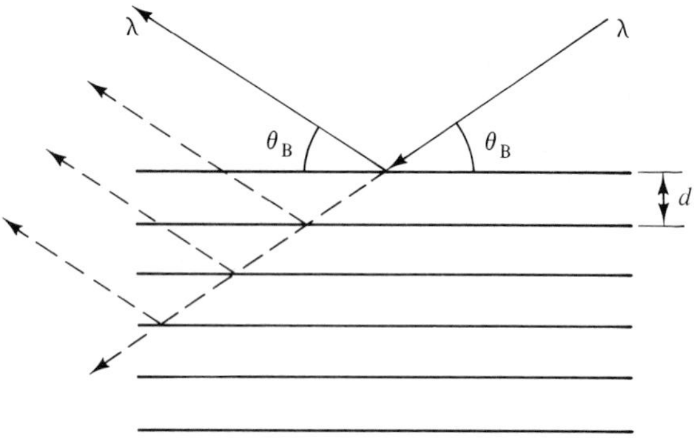

(*a*)

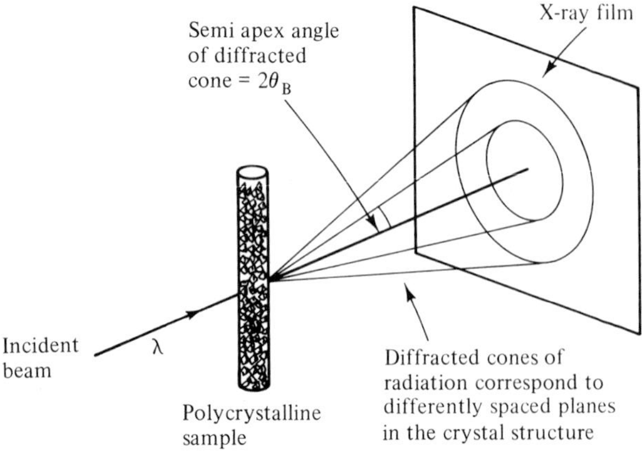

(*b*)

the radius of each giving θ_B and thus the 'd' spacing of the diffracting planes (Fig. 2.3(b)). In the case of powder diffractometry the one-dimensional pattern will correspond to a radial scan through these rings. In either case however, the polycrystalline (or 'powder') sample means that direct information about the relative orientation of the diffracting planes is lost. Where the sample is non-crystalline and possesses only short range positional order, the diffracting planes lose their identity and the resultant diffraction maxima are very diffuse, although their peak positions still contain information about mean molecular spacings.

Diffraction is particularly powerful where the sample has some degree of preferred orientation with respect to an external axis, and is thus very appropriate in the study of liquid crystalline materials. While it is impossible to explore the technique in depth here, the principles of interpretation of a transmission diffraction pattern are outlined with respect to a smectic C type sample in Fig. 2.4. A single pattern thus provides information about the layers, their orientation, and the perfection of their positional and orientational order. It also gives the orientational relationship between the molecules and the layers, the perfection of molecular packing (only short range order in this case) and the mean spacing of the molecules. Closer analysis would give further information about the exactness of the segregation into layers, and enable the quality of orientational and positional order to be quantified and the *order parameter*, *S*, determined. By convention, the axis on the diffraction pattern parallel to the rotation symmetry axis of the sample is known as the *meridian* (vertical on the figure), while that at right angles is called the *equator* (horizontal on figure).

Figure 2.5(a) shows a pattern from a nematic main-chain thermotropic polymer. The diffuse equatorial reflections again give information about the mean spacing of the molecules and the degree of alignment with the vertical axis, i.e. the order parameter. The horizontal (layer) lines represent the chemical periodicity of the molecules, and the fact that they are sharp in the vertical direction indicates that the molecules are homopolymers and in a straight, extended conformation. Figure 2.5(b) shows the pattern after crystallization. The equatorial maxima are now sharp and have split into two in this case to illustrate a unit cell with orthorhombic rather than hexagonal symmetry. Their arcing is much the same as in the nematic phase showing an orientational distribution which is not affected by the crystallization. The horizontal layer lines are now sampled as a series of arcs which demonstrates that the repeat units of the chains within the crystal are in longitudinal register with each other.

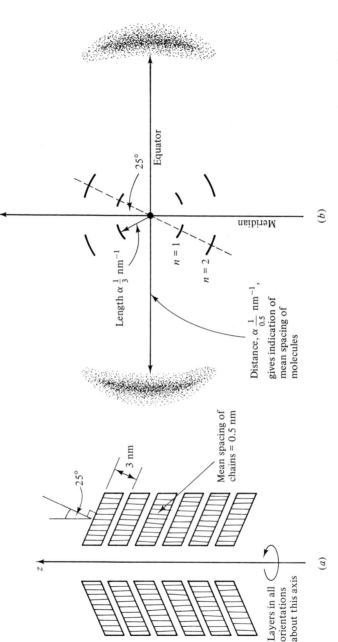

Fig. 2.4 A schematic diagram showing the relationship between a smectic C structure (a) and the various features of the resulting fibre diffraction pattern (b). The diffuse equatorial maxima in (b) show that the molecules are only packed to give short range order. The degree of arcing is related to the order parameter of the molecules. In contrast, the thin, concentric diffraction arcs from layers indicate long range positional order. The degree of arcing here indicates the perfection with which the layers are oriented.

Figure 2.6 is a powder diffractometer scan of a liquid crystalline polymer which has partially crystallized. The sharp peaks represent the crystalline component, while the diffuse maximum corresponds to the liquid crystalline matrix. Measurement of the relative areas under the diffuse and sharp parts of the scan can give the degree of crystallinity as indicated on the figure.

Fig. 2.5 A representation of a diffraction pattern from a nematic liquid crystal with the director vertical (*a*), compared with that from the same polymer after it has crystallized, (*b*).

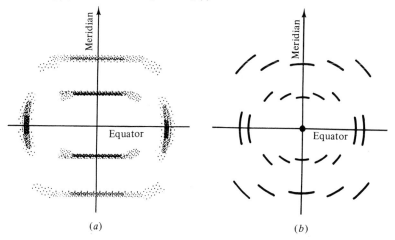

(*a*)   (*b*)

Fig. 2.6 A powder diffractometer scan of a thermotropic liquid crystalline polymer $\langle 32 \rangle$, in which the crystallinity is of the order of 20%. (Courtesy, Dr T. J. Lemmon.)

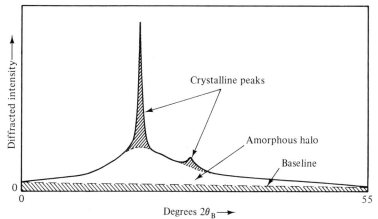

Director In a liquid crystal, there will be a direction towards which all the molecular orientations will be biased, at least locally. This property is the fundamental property of a liquid crystal, and the direction is known as the *director*. It is a unit vector, conventionally given the symbol **n**. Where the distribution has rotational symmetry, which is usually the case, the director is the axis of rotational symmetry. In more complicated distributions, such as those which might occur as a result of complex flow fields, the definition of the director is not as straightforward. It should, where possible, be identified with a rotation axis of the distribution, perhaps only a two-fold axis. In the normal case of uniaxial symmetry the *director* corresponds to one of the principal axes of the tensors which can be used to describe the properties of the material (see for example *optical indicatrix* under *Optical anisotropy*; in this case the director can be identified with one of the extinction directions observed in liquid crystals between crossed polars).

Distortions The three types of elastic orientational distortion which occur in liquid crystals, *splay*, *twist*, and *bend*, are illustrated in Fig. 2.7. The distortions are associated with a gradual change in the orientation of the director with position and do not in any way describe the distribution of molecular orientation about the director. In fact the theory of distortions is a continuum theory, and does not involve molecular parameters in any specific way.

Distortion can be simply understood by considering the director to point along the z axis, and by considering distortion as the sideways displacement of the head of the director vector, in the direction of either the x or y axes, as a function of translation within the material along either of the x, y or z axes. For the purposes of introducing the mathematics, we need to consider displacement of the head of the director in just the x direction. The coordinate systems marked on Fig. 2.7 are consistent with this statement.

Splay distortion, as depicted in Fig. 2.7(*a*), involves the rotation of the director, leading to a displacement of its head in the x direction, as a function of translation in the x direction. The splay distortion gradient can be written as:

$$(\partial n_x / \partial x)$$

or in vector notation, and more generally for any coordinate system as:

$$(\text{div } \mathbf{n})$$

Figure 2.7(*b*) shows the simplest example of a *twist* distortion, in which the z (and x) component of the director varies with its position along

Fig. 2.7 The three types of distortion of a nematic mesophase. In each of the examples: (*a*) splay, (*b*) twist and (*c*) bend, the director distortion is in the *x* direction. The result is that splay is $(\partial n_x/\partial x)$; twist, $(\partial n_x/\partial y)$; and bend, $(\partial n_x/\partial z)$.

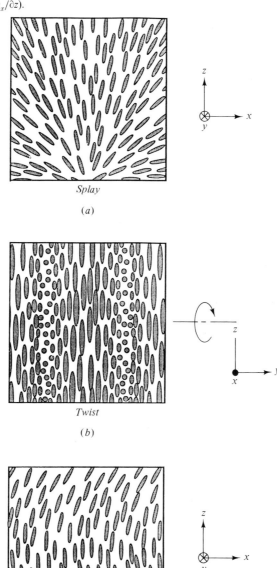

Splay

(*a*)

Twist

(*b*)

Bend

(*c*)

the y axis, but there is no component in the y direction because the rotation is occurring about the y axis. (Note that in Fig. 2.7(b) the length of the rods as depicted corresponds to the projection of the director onto the plane of the paper, so that the rods appear as circles where the director is normal to the page.) Considering the change in n_z, there will be a contribution of magnitude:

$$n_z \frac{\partial n_x}{\partial y}$$

Now turning to the more general case where twist is occurring about both x and y axes, this expression becomes

$$n_z \left(\frac{\partial n_x}{\partial y} - \frac{\partial n_y}{\partial x} \right)$$

For the most general case, the distortion can be written as:

n · curl n

where **curl n** is defined as:

$$\mathbf{curl\,n} = \left[\frac{\partial n_z}{\partial y} - \frac{\partial n_y}{\partial z}, \frac{\partial n_x}{\partial z} - \frac{\partial n_z}{\partial x}, \frac{\partial n_y}{\partial x} - \frac{\partial n_x}{\partial y} \right]$$

Bend distortion is shown in Fig. 2.7(c). Reasoning along the same lines, although the steps are slightly more complex, shows that bend distortion can be written as

n × curl n

The total free energy density of a liquid crystal with a general distortion, is the sum of the splay, twist and bend distortion components each multiplied by the stress level associated with the distortion. Each stress can be written in terms of an *elastic constant K* and the distortional strain:

i.e.   stress $= \frac{1}{2} K \times$ (strain)

Hence the free energy per unit volume (in vector notation) is:

$$F_{\text{distortion}} = \tfrac{1}{2} [K_1 (\text{div } \mathbf{n})^2 + K_2 (\mathbf{n} \cdot \mathbf{curl\,n})^2 + K_3 (\mathbf{n} \times \mathbf{curl\,n})^2]$$

The three elastic constants K_1, K_2, K_3, which apply to splay, twist and bend respectively are also known as the *Frank constants*. They have units of force, (i.e. N). In some texts they are ascribed double suffices e.g. K_{11}.

DSC (see **Differential scanning calorimetry**).
Elastic constants (see **Distortions**).

Elastic modulus (see **Mechanical properties**).
Electric and magnetic fields – molecular response Consider the mesogenic molecule ⟨1⟩. The nitrate group on one end tends to withdraw

$$CH_3-O-\!\!\!\raisebox{0pt}{⬡}\!\!\!\overset{\overset{O}{\|}}{C}-O-\!\!\!\raisebox{0pt}{⬡}\!\!\!\overset{\overset{O}{\|}}{C}-O-\!\!\!\raisebox{0pt}{⬡}\!\!\!-NO_2 \qquad \text{⟨1⟩}$$

electrons from the molecule with the result that it becomes negatively
charged; the methoxy group at the other end however, donates charge
to the molecule, so it is positively charged. The molecule is thus a small
electric dipole. The two ester groups are also dipoles with the double

Fig. 2.8 Illustration of alignment in an electric field due to the average
fixed dipole of the molecule, (a) and (b), and to an induced dipole (c). Both
mechanisms are often operative simultaneously.

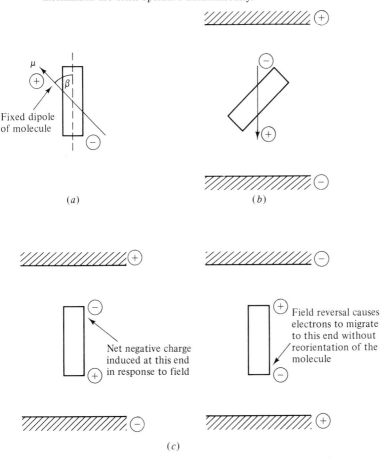

bonded oxygen negatively charged at the expense of the carbon atom. These dipoles subtend a large angle with the long axis of the rod-like molecule. The total dipole of the molecule, μ, will thus be inclined to the rod axis at an angle usually defined as β (Fig. 2.8(a)). Placed in an electric field the molecule can lower its energy by rotating so that the dipole axis lies parallel to the field with the positive end facing the negative electrode and vice versa (Fig. 2.8(b)). If the field reverses, as would occur if an alternating current is used, then the low energy state will only be maintained if the molecule is able to flip over to point in the opposite direction.

The movement of charge in response to the field is referred to as polarization. The constant of proportionality between the total electric field, represented by the vector $\mathbf{E}$, in a dielectric (non-conducting) material and the displacement, $\mathbf{D}$, is the permittivity tensor ε. Hence:

$$\mathbf{D} = \varepsilon\mathbf{E}$$

ε is also known as the dielectric tensor and is sometimes expressed as $\varepsilon = \varepsilon_0\varepsilon_r$ where ε_0 is the permittivity of free space and ε_r is the relative permittivity tensor. In an isotropic material ε_r is < 1 (see also the section on optical anisotropy).

As we are considering a nematic liquid crystalline state where the rod molecules have axial symmetry, it is convenient to define the dielectric anisotropy in terms of the components parallel to and normal to the axis as:

$$\Delta\varepsilon = \varepsilon_\parallel - \varepsilon_\perp$$

The dielectric anisotropy is given by:

$$\Delta\varepsilon = K\left[\frac{\mu^2}{2kT}(1 - 3\cos^2\beta)\,\mathcal{S}\right]$$

where $\mathcal{S}$ is the order parameter and K a constant for a given material. Note that $\Delta\varepsilon$ will be positive only for values of β less than $54.7°$.

A second contribution to $\Delta\varepsilon$ stems from the dipole induced in the molecule by the applied field (Fig. 2.8(c)). In a conjugated rod molecule such as $\langle 1 \rangle$ there is greater opportunity for electron transfer along the molecule than normal to it, so that the molecule has an anisotropic polarizability $\Delta\alpha_E$, which means that the energy in a field will be lowered when the rod axis is aligned with the field. In the case of an induced dipole however, the energy will be lowered whichever way up the molecule is. The alignment tendency is thus said to be quadrupolar rather than dipolar.

Where polarization is associated with the rotation of a permanent dipole, its contribution to the measured values of $\Delta\varepsilon$ depends on this rotation being possible at the frequency of the applied (AC) field. In the case of rod-like molecules, the frequency above which they cannot respond is often in the kHz range. The measurement of the dielectric constants as a function of frequency and temperature provide the basis of dielectric spectroscopy, which is a useful means of elucidating information about the structure and dynamic behaviour of these materials.

Molecule $\langle 1 \rangle$ will also respond to a *magnetic field* as it is diamagnetic (examples of paramagnetic molecules are rare). The effect is one of induced dipoles, and the molecular alignment which can be obtained is quadrupolar. Its strength is determined by the magnitude of the anisotropy in the diamagnetic susceptibility, written as:

$$\Delta\chi = \chi_{\parallel} - \chi_{\perp}$$

The induced magnetic dipole arises from changes in the orbital motion of electrons due to the applied magnetic field. The change in the electron motion itself creates a magnetic field which opposes the applied field and raises the energy of the system. It is thus favourable for molecules to align themselves so that the directions along which the electron motion is most affected lie parallel rather than normal to the field. The result is that the long axes of the molecular rods will align with the field, and, especially, the aromatic rings will tend to lie so that their planes are parallel to the field, not normal to it.

In describing briefly some of the principles underlying the orientation of organic molecules in electric or magnetic fields, it is easy to lose sight of the fact that liquid crystals in general align very effectively, while the same molecules in the isotropic phase would show a much more modest overall response. The reason is that the field orientation of a single molecule in the isotropic phase is opposed by random motion due to thermal energy. Put another way, the energy gained by a molecule orienting with respect to the applied field is often small compared to kT. However, when the molecule is in a mesophase, the random thermal motion of the rods is limited not only by the applied field but also by the strong quadrupolar alignment field of the neighbouring molecules.

Flory–Huggins parameter (see **Solution thermodynamics**).

Frank constants (see **Distortions**).

Friedelian classes There are three types of liquid crystal originally identified by Friedel in 1922 [2.2]. These are *nematic, cholesteric* and *smectic.*

The *nematic* liquid crystalline phase possesses long range orientational order but only short range positional order. The molecular

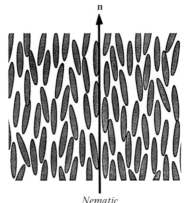

Nematic

(*a*)

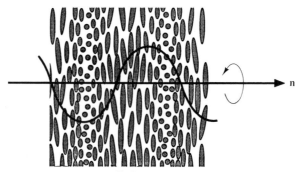

Cholesteric

(*b*)

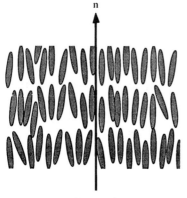

Smectic A   *Smectic C*

(*i*)   (*ii*)

(*c*)

(caption opposite)

organization is illustrated in Fig. 2.9(*a*). The rod-like molecules are aligned with respect to the vector known as the *director*, **n**. The quality of the alignment is not perfect, and is quantified by what is known as the *order parameter*, *S*. The symmetry of the system is no greater than the simple rod-like symmetry assumed for the molecules, the orientations of the rod-like units are distributed uniformly about the director which is therefore a full rotation axis, and **n** and −**n** are physically equivalent. Thus the molecular orientation possesses no 'sense' unlike, for example, in a ferroelectric material. In the light microscope, nematic liquid crystals show characteristic 'threaded' textures which is the origin of their name. (Greek $\nu\varepsilon\mu\alpha$ = thread.)

Examples of rod-like molecules which form nematic phases are *p*-azoxyanisole, known as PAA and drawn as molecule $\langle 2 \rangle$, and n-(*p*-methoxybenzylidene)-*p*-butylaniline, known as MBBA $\langle 3 \rangle$. The letters '*p*' in the molecular names stand for '*para*' which indicates the 1–4 connections at opposite points on the aromatic rings which preserves the linearity of the molecule, a key factor in the formation of a liquid crystalline phase.

$$CH_3{-}O{-}\langle\bigcirc\rangle{-}N{\overset{O}{\underset{\uparrow}{=}}}N{-}\langle\bigcirc\rangle{-}O{-}CH_3 \qquad <2>$$

$$CH_3{-}O{-}\langle\bigcirc\rangle{-}CH{=}N{-}\langle\bigcirc\rangle{-}(CH_2)_3{-}CH_3 \qquad <3>$$

The *cholesteric* state is illustrated in Fig. 2.9(*b*). It is equivalent to a nematic which has been twisted periodically about an axis perpendicular to the director, in fact the end result is the same as the twist type distortion of a nematic. However, the twist in a cholesteric arises spontaneously when the mesogenic molecules have a *chiral* nature. Chirality is associated with the presence of an asymmetric carbon atom, so that left and right hand versions of a molecule cannot be commuted by simple reorientation. Take for example the pair of molecules $\langle 4 \rangle$. No amount of rotation and/or inversion of the molecule on the left will make it equivalent to the one on the right. An asterisk * is conventionally used to designate the chiral centre. An example of a

---

Fig. 2.9 Representations of the molecular arrangements in the three Friedelian classes of liquid crystals: (*a*) nematic; (*b*) cholesteric (the twist axis is horizontal) and (*c*(*i*)) smectic A and (*c*(*ii*)) smectic C. Note that the structure of the cholesteric is similar to the imposition of a twist distortion on a nematic (Fig. 2.7(*b*)), although in the cholesteric case the twist is intrinsically stable.

group which is mesogenic as well as chiral is ⟨5⟩, the chiral centre is the carbon connected to methylene and ethane groups along the axis of the group, with the other two bonds being satisfied with a methyl group and a hydrogen atom. The steepness of the twist of cholesteric phases based on this molecule decreases as *m* increases and the chiral centre moves away from the mesogenic core of the molecule.

$$R_4-\overset{\overset{R_1}{|*}}{\underset{\underset{R_3}{|}}{C}}-R_2 \qquad R_4-\overset{\overset{R_1}{|*}}{\underset{\underset{R_2}{|}}{C}}-R_3 \qquad \langle 4\rangle$$

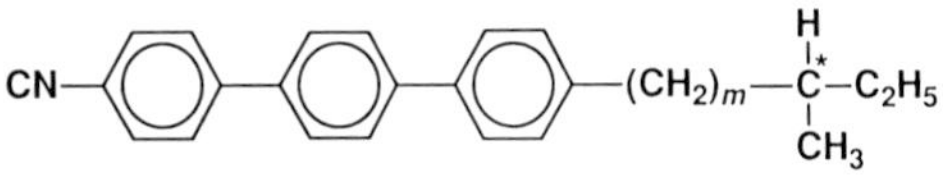

$$CN-\bigcirc-\bigcirc-\bigcirc-(CH_2)_m-\overset{\overset{H}{|*}}{\underset{\underset{CH_3}{|}}{C}}-C_2H_5 \qquad \langle 5\rangle$$

$$CH_3-(CH_2)_{12}-\overset{\overset{O}{||}}{C}-O- \qquad \langle 6\rangle$$

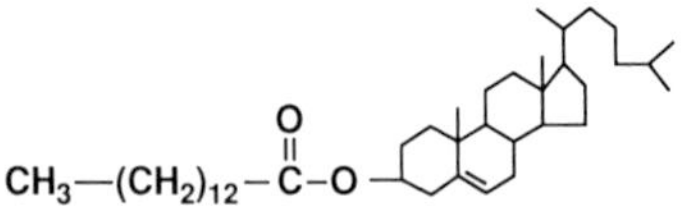

It follows that because there are two ways of arranging the substituents on the chiral centre, both left hand and right hand versions of the molecule can be made. If they are mixed in equal quantities (a so-called racemic mixture) then the rotatory effects will cancel, and a nematic rather than cholesteric liquid crystal will be the result. The class is called 'cholesteric' because this particular type of liquid crystalline organization was first observed in esters of cholesterol, such as ⟨6⟩.

Smectic liquid crystalline phases, are named because their basic layer structure gives them a soapy feel which is described by the Greek word σμεγμα. Indeed, many soap and detergent molecules exhibit lyotropic smectic phases in solution. These structures result from the *amphiphilic* nature of the molecule, which means that they have both a hydrophobic and a hydrophilic end. In solution, the molecules pack so that the hydrophobic ends come together, screening each other from the water. A similar structure is found in lipid bilayers, which are a fundamental constituent of cell walls.

There are several variants of smectics, but all are characterized by the possession of a *layered structure* on account of the segregation of the ends of the rod molecules onto common planes. The two most common variants are known as smectic A and smectic C and are shown in Fig. 2.9(c). In a *smectic A* phase the director lies along the layer normal.

The molecular packing within the layers is liquid-like and has no long range positional correlation. Likewise, there is no correlation between the lateral positions of the molecules in successive layers. Because of the disorder within the layers, the layers are not well defined, and in formal terms the smectic A can be described as a one-dimensional density wave. As in a nematic, the alignment of the molecules with the director is not perfect and is described by the order parameter.

Smectic C differs from smectic A in that the director of each layer is inclined at an angle ω to the layer normal, this angle being identical for all layers. If a chiral molecule is used a *chiral smectic C*, or C^*, structure may be formed in which the angle ω precesses about the layer normal. The structure consists of layers of tilted molecules twisted incrementally with respect to their neighbours. The molecules thus follow a helical path, the axis of which is the normal to the layers. Additional smectic variants are also known, in which there is increasing order. For instance in smectic B phases, the packing within the layers is on a two-dimensional lattice, but again there is no lateral correlation between layers.

The interplay between various levels of long range positional order and the orientational order basic to the liquid crystalline state means that a number of other smectic variants and sub-variants are possible and indeed have been observed. In their book on smectic liquid crystals, Gray and Goodby [2.3] identify the following types of smectic phase:

A, A_1, A_2, A_d, $\tilde{A}$
B, B_{hex}, B_2, B_A, B_C
C, C_1, C_2, $\tilde{C}$, C^*
D, E, F, F^*, G, G^*, H, H^*, I, J, K

Not all these phase designations are necessarily unique, and some are better described as crystals than as liquid crystals. The ideal classification has yet to be achieved. However, the list illustrates some of the richness of the field, and the definitive book [2.3] adds greatly to one's enjoyment. It remains to be seen whether many of the smectic sub-phases are relevant to liquid crystalline polymers, or whether other types of sub-classification will be found necessary.

In general, mesogenic molecules with terminal alkyl or alkoxy chains will tend to favour smectic phases, and the longer the terminal chains the more the smectic phase will be stabilized with respect to other *liquid crystalline* structures. Smectic phases also tend to be highly polymorphic, and in some instances a smectic liquid crystal will transform between a number of sub-classes during heating, the phases

26   *Terminology and concepts*

with less well-developed order being stable at the higher temperatures. An example of such behaviour is provided by the molecule ⟨7⟩, which shows the following transitions [2.4]:

$$\text{Crystal} \xrightarrow{15°} S_G \xrightarrow{35°} S_B \xrightarrow{62°} S_A \xrightarrow{77°} N \xrightarrow{81°C} \text{Isotropic}$$

$$C_6H_{13}-O-\langle\bigcirc\rangle-CH{=}N-\langle\bigcirc\rangle-O-C_6H_{13} \qquad \langle 7 \rangle$$

Glass transition Most liquids crystallize on cooling. However, where either a lack of molecular symmetry or a sufficiently high cooling rate is able to inhibit crystallization, solidification is associated with the formation of a glass. In many respects a glass is simply a frozen sample of the liquid from which it formed. The molecules do not show any long range order and an X-ray diffraction pattern will be qualitatively the same as that of the liquid, showing none of the sharp peaks characteristic of a crystal.

In a liquid, fluidity is associated with the freedom of molecules to move with respect to each other on a suitably short time scale. The density of the liquid state, often greater than 90% of that of the corresponding crystal (or in the case of water greater than 100%!), means that the molecular movement must be cooperative. As the liquid is cooled, if we assume that crystallization is inhibited, not only will the average thermal energy available to the molecules be reduced, but also densification occurs. Both these effects will reduce the possibility of further large scale molecular motion taking place, and the densification itself requires cooperative motion to permit it to proceed. Ultimately these long range motions will be completely frozen out.

The situation is represented schematically in Fig. 2.10. The solid line is the equilibrium relationship between specific volume (volume of unit mass of material) and temperature. The figures along the top represent estimates of the times (in seconds) which are required for the necessary molecular readjustments to maintain densification within each of the temperature bands which might be around 10 °C wide, as in this example. Imagine the liquid to be quenched so that it cools across each temperature band in 10^{-3} s. The specific volume will follow the equilibrium line until it reaches the 1 ($= 10°$) s band. Within this band there will be insufficient time for the full densification and, on moving into lower temperature bands, the specific volume will become frozen at a value characteristic of the cooling rate as shown by the upper dashed line. In real systems, the specific volume continues to reduce at lower temperatures but at a much slower rate. This continued, but decreased, shrinkage is associated with the reduction in molecular vibration amplitude, and is not shown on the diagram. The temperature at which the specific volume plot changes slope, is known as the *glass transition temperature* or T_g, and is marked T_g^i on the diagram.

If the liquid is cooled at a much slower rate, say 10^3 s per temperature band (as might occur if the liquid was being cooled in a dilatometer in order to measure the specific volume), the volume will follow the equilibrium line until a lower temperature is reached, as is represented by the lower dashed curve. The glass transition temperature is thus considerably lower, at T_g^{ii}.

The fact that the glass transition temperature depends on the time scale over which it is measured underlines the fact that it is *not a*

Fig. 2.10 Schematic plot of specific volume against temperature to illustrate the glass transition temperature. The vertical bands suggest the times which might be necessary for the coordinated rearrangement of molecules in the different temperature ranges. T_g^i is the glass transition temperature for a fast cooling rate, T_g^{ii} that for a slow cool. The change in the component of specific volume associated with the dependence of molecular vibration amplitude on temperature leads to a further (reduced) contraction below T_g, which is not shown.

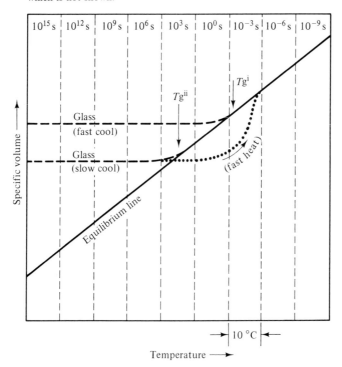

thermodynamic transition. A glass is not an equilibrium phase and any attempt to identify the transition with a thermodynamic order is misleading. The value of T_g will also depend on heating rate. Take for example the slower cooled glass: if it is heated very rapidly, the glass molecules will not have sufficient time to readjust their positions and will expand until they reach a temperature greater than the T_g on cooling. The plot would thus overshoot before it returns to the equilibrium line (dotted). Further details are contained in most general polymer texts, e.g. [2.5].

The glass transition is relatively more significant in polymers than in small-molecule substances. Not only do the polymer chains provide good opportunities to frustrate crystallization through stereoisomerism or random copolymerization, but the connectivity inherent in the chain structure means that greater molecular cooperation is necessary for internal readjustment.

Homeotropic texture (see **Surface orientation**).

Homogeneous (planar) texture (see **Surface orientation**).

Homopolymer A *homopolymer* is a polymer built up from a series of identical repeating units. The term is used to distinguish such a polymer from a *copolymer*.

Lyotropic liquid crystals Lyotropic liquid crystalline phases contain non-mesogenic solvent molecules in addition to the mesogenic ones. The term is used to distinguish the systems from *thermotropic* liquid crystals which consist of mesogenic molecules alone. Lyotropic systems are particularly significant in liquid crystalline polymers, as the addition of the solvent is an important means of reducing crystalline melting points to manageable levels.

Magnetic fields (see **Electric and magnetic fields**).

Mechanical properties Stress describes force per unit area. Axially opposed forces provide tensile or compressive fields, the cross sectional area over which they act being normal to the axis (Fig. 2.11(a)). A shear stress is the result of two opposing forces which are offset with respect to each other. The area on which the forces act is parallel to the force axis and normal to the direction of offset (Fig. 2.11(b)). Normal stresses are usually designated σ, and shear stresses, τ. A hydrostatic stress acts equally in all directions, either as a tension or compression, and it is possible to resolve a normal stress into hydrostatic and shear components, the latter acting on planes at $\pm 45°$ to the axis. The units of stress are newtons per square metre ($N\ m^{-2}$) which are also known as pascals (Pa).

As stress relates two vectors, the force and the area on which its acts, it is a second rank tensor and requires two suffixes to identify it. Where the three orthogonal directions are designated 1, 2 and 3, and the orientation of a plane is indicated by the direction of its normal, then a normal stress can be written:

$$\sigma_{11} = F_1/A_1$$

and similarly for σ_{22} and σ_{33}.

For a shear stress, where the forces are acting along the 1 axis, but are offset in the 2 direction and act on planes normal to this direction, then:

$$\tau = F_1/A_2 = \sigma_{12}$$

Other shear stresses can be similarly designated, σ_{13} and σ_{23}. It is interesting to note that a shear stress produces a force couple, which in isolation would cause the material to rotate in space. Thus for any static scenario, the shear stress σ_{12} must be exactly opposed by the shear stress σ_{21}, and σ_{23} by σ_{32} etc.

There are various critical stresses which are used to describe materials behaviour. The yield stress is the stress necessary to induce plastic (permanent) deformation in a material, whereas the fracture

Fig. 2.11 Geometry of (*a*) a normal stress (*b*) a shear stress.

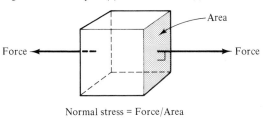

Normal stress = Force/Area

(*a*)

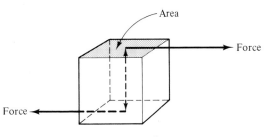

Shear stress = Force/Area

(*b*)

stress or 'ultimate' strength is the critical stress for fracture. Where a material deforms prior to the critical strength being reached, whether that is elastic (recoverable) deformation prior to the yield stress or elastic and plastic (non-recoverable) deformation prior to the fracture stress, the cross sectional area used in calculating the stress will have changed from its original (zero deformation) value. If the original cross sectional area, A_0, is used in calculating the stress, the stress is referred to as a nominal stress, and usually designated σ_n; if the cross section used is that at the actual strain reached then the stress is called a true stress, σ_t.

Where, for example, there is considerable deformation of a polymer prior to fracture, the true fracture stress can be several times greater than the nominal fracture stress.

Strain describes displacement expressed as a fraction of the original dimensions of a material. As with stress there are normal, shear and hydrostatic strains. Strain can be expressed as the change in length per unit original length, or as the ratio of new length to original length, or as small increments of length change each referred to the 'original' length just prior to the increment; the increments being integrated over the total deformation to give the so-called 'true strain'. In the case of shear strain, the displacement is measured in the directions of the shearing forces, but the 'original' length is the offset between the displacement vectors.

Normal strains can thus be defined as a percentage extension:

$$\varepsilon = 100 \times (l - l_0)/l_0$$

extension ratios as:

$$\varepsilon_r = l/l_0$$

and true strains as:

$$\varepsilon_t = \int_{l_0}^{l_{max}} \frac{\delta l}{l} = \ln(l_{max}/l_0)$$

Continuing the designations used for stress, normal strains are written ε_{11}, ε_{22} etc, and shear strains ε_{12} etc. Like stress, strain is a tensor.

It can be shown that the reduction in cross sectional area with normal strain, expressed as the ratio A/A_0, is equal to l_0/l for an incompressible material. Hence $A/A_0 = \varepsilon_r^{-1}$, so that the true stress, σ_t is related to the nominal stress, σ_n as follows:

$$\sigma_t = \varepsilon_r \sigma_n$$

As with stress, there will be critical strains corresponding to different stages of the response of the material to stress. As strain is a ratio of lengths it has no units. The *elastic modulus* of a material is a measurement of its stiffness. In other words it is the stress necessary to produce a unit of elastic strain. In defining the modulus, the unit of strain is usually taken to be 100% extension, or a doubling in length. The units of elastic modulus are thus those of stress, newtons per square metre $(N\,m^{-2})$ or pascals (Pa). However, the elastic modulus can only be viewed as a constant over small strains. In cases where the polymer yields or fractures at 10% strain or so, the slope of the elastic portion of the stress strain curve is normally taken to give the modulus. Tensile or compressive moduli are the ratios of normal stresses to normal strains, and correspondingly, shear moduli are the ratios of shear stresses and strains.

Specific values of strength and stiffness are often used when comparing different materials for applications such as in aerospace where weight is a penalty. These values are derived by dividing the mechanical parameter, expressed as stress, by the specific gravity (SG) of the material which is its density normalized to that of water. The units thus remain those of stress.

Hence specific stress is usually written as: $GPa\,(SG)^{-1}$ or in some cases as $GPa\,(kg\,m^{-3})^{-1}$.

Another set of specific units used in fibre science is the force in grams per denier $(g\,den^{-1})$ or more recently $N\,tex^{-1}$. The denier unit is defined as the mass in grams of 9000 metres of yarn. Hence:

$$\text{wt in g (den)} \equiv \rho(g\,m^{-3}) \times A(m^2) \times 9000$$
$$\equiv 10^6 \times \rho(g\,cm^{-3}) \times A(m^2) \times 9000$$

The applied force will be in grams force (g f), now:

$$\text{Force (g f)} \equiv \frac{\text{Force (N)}}{981 \times 10^{-5}}$$

$$\therefore \frac{\text{Force (g f)}}{\text{wt. in g (den)}} \equiv \frac{\text{Force (N)}}{10^6 \times \rho(g\,cm^{-3}) \times A(m^2) \times 9000 \times 981 \times 10^{-5}}$$

$$\equiv \frac{11.33 \times 10^{-9} \times \text{Stress (Pa)}}{\rho(g\,cm^{-3})}$$

$$\equiv 11.33 \times \frac{\text{Stress (GPa)}}{\text{SG}}$$

Hence stress (or modulus) in GPa can be converted to $g\,den^{-1}$ by first converting to specific stress by dividing by the SG (density in g cm^{-3}) of the material, and then converting to g den^{-1} by multiplying by 11.33. Another fibre unit, the tex, is defined as the weight in grams of 1 kilometre of fibre. Hence, one would convert denier to tex by dividing by 9, and the stress unit of N tex^{-1} is exactly equivalent to the specific stress in GPa (SG)$^{-1}$.

Neither g den^{-1} nor N tex^{-1} are used in this book, however several key reference texts use them liberally.

The descriptive account above serves to remind the reader of the definition of some of the more important terms relevant to this text. However, stress, strain and elastic modulus (for example) are all tensorial and there is a complete formalism for their manipulation. For a more detailed approach, reference should be made to other sources such as [2.6].

Mesogenic (see **Mesophase**).

Mesomorphic (see **Mesophase**).

Mesophase *Meso*, from the Greek μησος meaning middle, was the term introduced by Friedel to describe liquid crystalline attributes. It is possible to criticize Lehmann's original naming of 'liquid crystals' on the basis that such structures are definitely not crystalline, and sometimes not liquid. It was suggested therefore that they should be called *mesophases* as they were stable in the region between crystal and liquid phases. Other words also stem from the *meso* root.

Mesomorphic is used to describe the structural attributes of a liquid crystalline phase, while *mesogenic* is applied to molecular species which are capable of forming liquid crystalline phases. In particular, repeat units which are sufficiently rigid and straight that they will polymerize to form a liquid crystalline polymer are referred to as mesogenic units. The words mesophase and mesomorphic are used synonymously with liquid crystal phase and liquid crystalline in this text.

Molecular weight Unless special polymerization procedures are used and extraordinary care taken, the molecules of a polymer have a considerable range of lengths and thus molecular weights. The distribution of molecular weights means that care has to be exercised in assigning a characteristic value. Two types of average are in common use, the *number average molecular weight* is given by:

$$M_n = \sum_i X_i M_i$$

where X_i is the number fraction of molecules present with molecular weight M_i.

It is possible that a considerable proportion of the polymer may consist of comparatively few molecules of extremely high molecular weight. Such molecules can have a very large influence on properties such as viscosity, and they are better represented in a molecular weight average which is the average of the weight distribution of the individual molecules. The *weight average molecular weight* is given by:

$$M_w = \sum_i W_i M_i$$

where W_i is the weight fraction of molecules of molecular weight M_i.

The ratio M_w/M_n gives some indication of the breadth of the molecular weight distribution, and is sometimes known as the *polydispersity*. For polymers formed by condensation polymerization, this ratio is found to be close to the theoretical prediction of 2.

The number of monomer units in a polymer chain is known as the *degree of polymerization* (*DP*), and is more generally taken to imply the number average DP, although this is not often spelt out.

Nematic (see **Friedelian classes**).

Newtonian flow (see **Viscosity**).

Nuclear magnetic resonance (NMR) Protons (hydrogen nuclei, ^{1}H) and certain other nuclei (most importantly for polymers ^{13}C) possess a spin. When those nuclei that have spin $\frac{1}{2}$ (such as protons and fluorine) are placed in a strong magnetic field (typically of the order of a tesla or more), they may occupy one of two states, separated in energy, according to whether their spins align parallel or antiparallel to the field. Under certain circumstances it is possible to detect slight chemical shifts arising from shifting of the energy levels of the nuclei due to the differing environments they find themselves in because of the detailed interactions of the nuclei with neighbouring chemical bonds. Such chemical shifts can only be detected if other interactions such as dipole–dipole interactions are unimportant; this can be achieved by application of certain pulse sequences or because of molecular motion. This kind of chemical shift is very important when examining the make-up of molecules, and in particular their tacticity.

For liquid crystals it is possible, albeit difficult, to measure the *order parameter S* from ^{1}H NMR. This can be done because pairs of protons, for instance on a benzene ring, will give rise to a doublet (known as the Pake doublet) in the spectrum arising from intra-molecular dipole interactions between the two protons. Unfortunately this doublet is often obscured by other intermolecular interactions. However, for a solid oriented sample the change in splitting of the doublet as the sample is rotated in a magnetic field can sometimes be detected, and this can be used to determine *S*.

A more powerful technique for the study of orientation in liquid crystalline polymers uses deuterium, ^{2}H, NMR. This is a more successful method because the much smaller nuclear moment of ^{2}H compared with ^{1}H reduces the magnitude of the magnetic dipole interactions, and it is the quadrupolar interaction which is used. The spectra from ^{2}H NMR are much cleaner and simpler than those from ^{1}H NMR, and thus interpretation is easier. In the case of ^{2}H NMR, a doublet spectrum is observed which depends on the interaction with the gradient of the electric field. For aliphatic compounds this will be collinear with the C—^{2}H bond, but this collinearity is only approximate for aromatic compounds. In the liquid state, where molecular motions are rapid compared with the measurement time (which determines the spectral width), the C—^{2}H bond vectors on average lie along the liquid crystal director, and thus an analysis of the shapes of the spectral lines which show the quadrupolar splitting will give a measure of the order parameter. While this approach may be readily applicable to small-molecule liquid crystals, various additional factors arise when polymeric systems are considered. Firstly the molecules are often very heterogeneous, as for example in the case of side-chain material, so that one must ask the question as to which part of the molecule the measured order parameter applies. Secondly, they are often quenchable to the glassy state in which case the molecular motion no longer permits the assumption that the C—^{2}H vectors average along the director. Furthermore, there will be a distribution of mobilities within the molecule, which at particular temperatures can further complicate the analysis. The strength of the method is that these additional factors can be used to learn much more about the liquid crystalline state.

The selective substitution of hydrogen atoms with deuterium in specially synthesized molecules (called labelling), provides a very powerful method of analysing both the local ordering and the motions associated with particular bonds or groups. In the rapid motion limit, labelling enables the order parameter of a particular part of the molecule to be determined, for example a mesogenic side group, or a flexible spacer. Observations of the changes to the spectrum as the rapid motion averaging is lost on cooling can enable the dynamic characteristics of a specified part of the molecule to be determined also. The determination of the structure in a liquid crystalline polymer glass usually involves the calculation of line shapes for various possible models. These must take into account the motions which remain in the glass but are gradually frozen with further decreases in temperature.

The fact that there is a natural concentration of the ^{13}C isotope in organic material, means that the technique does not require specially synthesized molecules. The different chemical shifts associated with the different carbon atoms in the molecule can be distinguished, if the anisotropic components of the chemical shifts are averaged out by rapidly spinning the sample at what is called the 'magic angle': ($\cos^{-1}$ $(1/\sqrt{3}) = 54.7°$). It is possible to determine information about both the orientational order and the structural environment of the ^{13}C—H bonds by using what is called two-dimensional magic angle spinning NMR. The information is displayed with the chemical structure in one dimension and the order in the other. Only in macroscopically oriented samples do side bands appear in the order dimension. A more detailed review of the application of NMR methods to liquid crystalline polymers is given by Boeffel and Spiess [2.7].

Optical anisotropy One of the fundamental properties of liquid crystals (as indeed of crystals) is that they are optically anisotropic. This means that the propagation of a ray of light through the medium depends on its orientation. In general, optical properties of a transparent substance depend on the relationship between three fundamental vectors:

D is the displacement vector describing the light within the material and is parallel to the direction of polarization of the incident light, **E** is the electric field vector within the material and is modified by the polarization of the molecules induced by the incident beam, **k** is the wave vector which is normal to the propagating wave fronts and has a magnitude equal to $2\pi/\lambda$, where λ is the wavelength.

Restrictions on the interrelationship of these vectors follow from Maxwell's equations. These restrictions are that the **D** vector is always normal to **k**, and that the **E** vector must lie in the same plane as **D** and **k**.

The displacement vector **D** is related to **E** through the permittivity tensor ε.

Hence,

$$\mathbf{D} = \varepsilon\mathbf{E}$$

or as **E** is the total resultant electric vector in the material, and **D** the displacement characteristic of the light beam, it is normal to rewrite the equation:

$$\mathbf{E} = \varepsilon^{-1}\mathbf{D}$$

where ε^{-1} signifies the inversion of the tensor.

The algebra implied by a tensor can be illustrated by what is known as a representative surface. It is a three-dimensional ellipsoid based on three principal axes, aligned with the principal axes of the tensor. If the

(a)

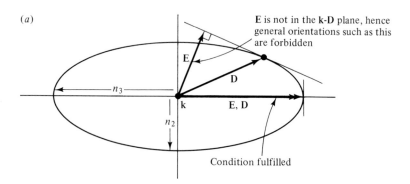

E is not in the **k-D** plane, hence general orientations such as this are forbidden

Condition fulfilled

(b)

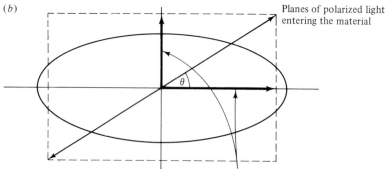

Planes of polarized light entering the material

The allowed components within the material which get out of phase because they experience different refractive indices, n_3 and n_2.

(c)

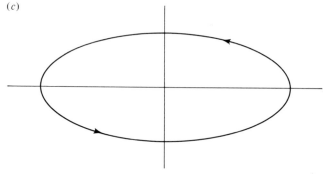

(caption opposite)

values of the tensor along its principal axes are ε_1^{-1}, ε_2^{-1}, ε_3^{-1}, as in this case, then the lengths of the principal axes of the ellipsoid will be proportional to the reciprocal of the square roots of the three quantities, that is $\sqrt{\varepsilon_1}$, $\sqrt{\varepsilon_2}$, $\sqrt{\varepsilon_3}$. Now because a principal refractive index, n, is proportional to $\sqrt{\varepsilon}$, the lengths of the principal axes of the ellipsoidal representation surface are proportional to the three principal refractive indices, n_1, n_2, and n_3. For the case of optical properties this representation surface is known as the *optical indicatrix*.

In the case of a sample with uniaxial rotational symmetry such as a nematic, with the symmetry axis set along z, then $n_1 = n_2$. Taking a cross section of the ellipsoid in the 2,3 plane, with the wave vector **k** normal to it and thus parallel to n_1 (Fig. 2.12(*a*)), and given that the orientation of the vector **E** corresponding to a particular **D** is the normal to the tangent plane at the point where **D** intersects the surface, it can be seen that the only two possible conditions which the Maxwell restrictions permit are those where **E** and **D** are both collinear with each other and either the major or the minor axis of the elliptical section. Hence, the only states of polarization permitted within the optically anisotropic medium are those parallel to the major and minor axes of the elliptical section of the optical indicatrix. The situation is a little more complicated when **k** is not parallel to any of the principal axes of the indicatrix, but this will not be developed further here.

When plane polarized light enters the material parallel to either of the principal axes of the section, it is not disturbed as it propagates through the medium. Hence, it will be completely stopped by a polaroid placed at right angles to the original polarization. Thus, a liquid crystal will appear dark between crossed polars when its director (or the projection of the director onto the viewing plane) is parallel to either the polarizer or analyser.

Plane polarized light which makes some angle θ with the principal axes of the section will propagate as two components parallel to each of the principal axes (Fig. 2.12(*b*)). They can be considered separately

---

Fig. 2.12 (*a*) Elliptical section through a uniaxial optical indicatrix which contains the unique symmetry axis (horizontal). The diagram shows that when the displacement vector **D** is in a general position (in this plane) the direction of the electric vector, **E**, is not in the **k**–**D** plane, and thus not permitted (**k** is parallel to the normal to the page). The condition, in this example, is only fulfilled when **E** and **D** are collinear with a principal axis of the elliptical section. (*b*) The resolution of plane polarized light into the two permitted components, each parallel to a principal axis of the section and each experiencing a different refractive index. (*c*) Illustration of elliptically polarized light produced by the recombination of orthogonal components of unequal amplitudes with a phase difference of $\pi/2$.

within the anisotropic material, and as they will each be experiencing a different refractive index they will gradually get out of phase with each other. Two orthogonal components, with a general phase difference between them of ϕ will recombine to give what is known as elliptically polarized light (Fig. 2.12(c)) which cannot be completely stopped by the analysing polaroid whatever its angular setting. The contrast will thus appear bright, except in the special case where the thickness of the sample leads to a phase difference which is exactly $2\pi n$, when the sample will appear dark, at least in monochromatic light, whatever the setting of the crossed polars.

There are many other aspects to the understanding of optical properties of anisotropic materials, and only the simplest case of a uniaxial material with the director lying in the plane of the specimen has been followed through here. The interested reader is referred to a standard optical text such as Chapter 5 of [2.8].

Optical indicatrix (see **Optical anisotropy**).

Order parameter Molecular orientation is of central significance in the fields of both polymer science and liquid crystal science. When the two subjects came together in the study of liquid crystal polymers, they were both well equipped in the methods of quantifying the degree to which molecular axes were aligned with a particular reference direction. In polymers this direction may typically be the draw axis of a fibre, while for liquid crystals it would more likely be the direction of an electric field applied to align the molecules. At first sight the parameters developed to describe the level of preferred orientation appear quite different. Polymer scientists refer to the orientation function, or specifically the Hermans orientation function, and designate this quantity P_2. On the other hand, the equivalent liquid crystal term is *order parameter*, and is designated by the symbol, S (sometimes upper case, sometimes lower). However, both parameters are in fact identically derived, and completely equivalent.

Consider a liquid crystalline phase in which the rod-like molecules are aligned with respect to the *director* which is the symmetry axis of the orientation distribution. The probability distribution of the long axes of the molecules can be represented by an ellipsoidal surface which has rotational symmetry about the director (Fig. 2.13(a)). The probability of a molecule having a particular orientation is proportional to distance from the origin to the surface along a line in that direction. The greater the quality of molecular alignment with the director, the longer and thinner will be the ellipsoid, so that in the limit of perfect orientation it will reduce to a single line.

The next step is to find a convenient method of characterizing the

Fig. 2.13 (*a*) Surface representing an orientation probability function, $\rho(\alpha)$, for uniaxial distribution. (*b*) Polar plot showing the form of some of the even components of the series $P_n(\cos\alpha)$. (*c*) A diagram envisaging an arrangement of liquid crystalline domains to illustrate the distinction between local directors **n** and a global director, **N**.

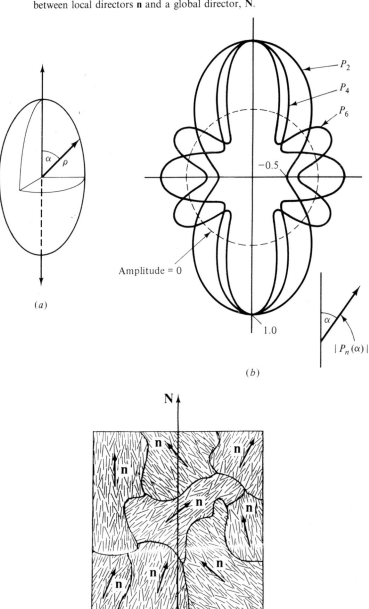

(*a*)

(*b*)

(*c*)

shape and axial ratio of the ellipsoidal probability distribution. It is possible to analyse the ellipsoidal shape into a series of *spherical harmonics* much as one might analyse the form of a simple periodic function into Fourier components. The component functions must be orthogonal, which means that a change in one does not consequently change any other, and that the product of any two will integrate to give zero (if they are different) or unity (if they are the same). For a Fourier series the components are simply sine or cosine functions with wavelengths exact fractions of the period of the periodic function. The orthogonal functions which will combine to reconstitute the ellipsoidal surface are more complicated. They are in fact spherical harmonic components, and are usually referred to as the even members of a series $P_n(\cos \alpha)$; odd components are not required for a shape which shows mirror symmetry about a plane normal to the rotational symmetry axis, all of them being identically zero. The components are defined in terms of $\cos \alpha$ where α is the angle between the rod-like molecules and the director as defined in Fig. 2.13(*a*). The spherical harmonic functions have the form:

$$P_2(\cos \alpha) = \tfrac{1}{2}(3 \cos^2 \alpha - 1)$$
$$P_4(\cos \alpha) = \tfrac{1}{8}(35 \cos^4 \alpha - 30 \cos^2 \alpha + 3)$$
$$P_6(\cos \alpha) = \tfrac{1}{6}(231 \cos^6 \alpha - 315 \cos^4 \alpha + 105 \cos^2 \alpha - 5) \text{ etc.}$$

These functions are illustrated as polar plots in Fig. 2.13(*b*), and the addition of the component members of the series each with an amplitude given by the appropriate P_n, will produce any ellipsoidal shape described by the function $\rho(\alpha)$. It follows that the shape is characterized by the series of amplitudes, $\langle P_2(\cos \alpha)\rangle$, $\langle P_4(\cos \alpha)\rangle$, $\langle P_6(\cos \alpha)\rangle$ etc. and that:

$$\rho(\alpha) = \sum_{n=0}^{\infty} (n + \tfrac{1}{2}) \langle P_n(\cos \alpha)\rangle P_n(\cos \alpha)$$

where the term $(n + \tfrac{1}{2})$ is a weighting factor.

In liquid crystal science, the order parameter $\mathcal{S}$, which is the same as P_2, is almost universally used and the higher harmonics are usually ignored. The order parameter is thus a very useful guide as to the quality of the orientation, but it gives little indication of the actual form of the orientation probability function, $\rho(\alpha)$, or in other words the detailed shape of the ellipsoidal surface. *The term order parameter rather than the orientation function ($\mathcal{S}$ rather than P_2) is used throughout this text.*

The properties of $\mathcal{S}$ (and indeed P_2) are that it has a value of 1 for perfect alignment, 0 for random orientation and $-(\tfrac{1}{2})$ for orientation in a plane, i.e. all the molecules lying in a plane at right angle to the director but without any preferred orientation within the plane itself.

It is important to appreciate that although the order parameter measure with respect to an external director, about which the molecules are symmetrically disposed, is an unequivocal number, it is often possible to recognize a hierarchy of directors. A simple example is shown in Fig. 2.13(c). Here a domain structure is envisaged. In each domain, the molecules are oriented with respect to the director, **n**, characterizing the domain. The domains themselves are then oriented with respect to the global director **N**. There are thus three order parameters: S^A, S^B and S^C, the first characterizing the quality of molecular orientation within each domain with respect to **n** for that domain, the second the orientation distribution of the domain directors, **n**, about the global director, **N**, and the third the orientation of the molecules with respect to **N**. While the separation between the two regimes is very clear in this example, it is apparent situations can arise in which any measured value of the orientation parameter could be a continuous function of the volume sampled.

One great advantage of using the order parameter, which corresponds to an amplitude of a harmonic component of the full orientation function, is that when one orientation distribution is convoluted with another, as in the example above, there is a simple relationship between the three values of S. The order parameter of the molecules with respect to the eternal 'global' director, is given by the simple convolution of the two parameters. In this case therefore:

$$S^C = S^A \cdot S^B$$

Orientation function (see **Order parameter**).
P_2 (see **Order parameter**).
Persistence length (see **Chain dimensions**).
Planar orientation (see **Surface orientation**).
Polymer synthesis Progress in the field of liquid crystalline polymers is dependent on the skill of the synthetic chemist. The development of new mesogenic materials for particular applications is an example of engineering at the molecular level. Not only is it vital to be able to synthesize and purify a molecule in the first place, but the economics of the process must be kept in view if it is to be scaled up for commercialization. Three types of polymerization reaction cover most of the syntheses of liquid crystalline polymers.

The *main-chain thermotropic polymers* developed for moulding applications, are almost entirely polyesters, or polyamides. They are polymerized using step-growth such as condensation reactions in which a small molecule is eliminated as each new monomer is added to the chain. There are three important reactions for making polyesters, all of which involve a carboxylic acid or a simple derivative of that acid, and these will be illustrated in the context of the synthesis of a random aromatic copolyester, based on a difunctional alcohol and a difunctional acid or derivative. The same principles apply for monomers with an acid group at one end and an alcohol group at the other. The symbol 'Ar' stands for *any* type of aromatic ring such as phenylene, naphthalene, bi-phenyl etc.

(i) Acid + alcohol. This reaction occurs in the presence of a phosphorous compound as a dehydrating agent, and a chlorocarbon.

$$HOOCArCOOH + HOArOH + (C_6H_5)_3P + C_2Cl_6$$
$$\rightarrow HOOCArCOOArOH + (C_6H_5)_3PO + C_2Cl_4 + 2HCl$$

(ii) Acid chloride + alcohol. This reaction is also known as the Schotten–Baumann reaction, and may be carried out very simply either in solution at elevated temperature, by interfacial condensation at room temperature, or as a melt reaction.

$$ClOCArCOCl + HOArOH \rightarrow ClOCArCOOArOH + HCl$$

(iii) Acid + ester. An example would be the acidolysis reaction of a diacetate ester with the diacid. This reaction is particularly easy to control and leads to the liberation of acetic acid which has to be pumped off in the later stages to achieve usefully high molecular weights.

$$HOOCArCOOH + CH_3COOArOOCCH_3$$
$$\rightarrow HOOCArCOOArOOCCH_3 + CH_3COOH$$

The acidolysis reaction has also been applied commercially to the preparation of random copolyesters by the insertion of monomers into high molecular weight polymers. By far the most important example involves the reaction of an acetate ester of hydroxybenzoic acid with a high molecular weight poly(ethyleneterephthalate) (PET).

$$[-OCH_2CH_2OOCArCO-]_n + qCH_3COOArCOOH$$
$$\rightarrow [OCH_2CH_2OOCArCO-]_{n/(n+q)}[-OArCO-]_{q/(n+q)}$$
$$+ qCH_3COOH$$

The PET polymer chain has the benzoic acid units substituted at random along it. Hence, its aromaticity is increased which leads to the formation of a mesophase, and it becomes a random copolymer which inhibits crystallization and reduces the melting point. The molecule is ⟨33⟩.

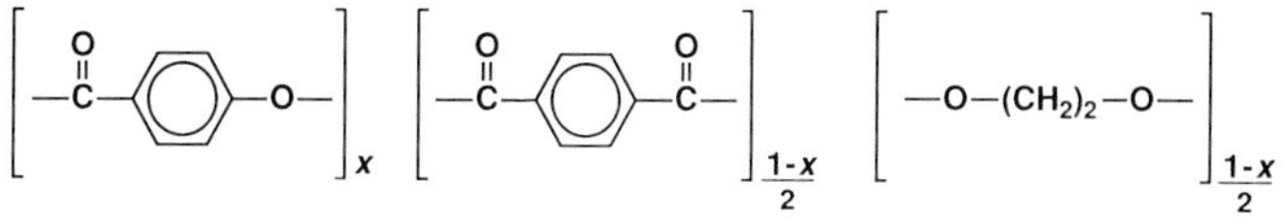

$$\left[-\overset{O}{\underset{\|}{C}}-\hspace{-4pt}\bigcirc\hspace{-4pt}-O- \right]_x \left[-\overset{O}{\underset{\|}{C}}-\hspace{-4pt}\bigcirc\hspace{-4pt}-\overset{O}{\underset{\|}{C}}- \right]_{\frac{1-x}{2}} \left[-O-(CH_2)_2-O- \right]_{\frac{1-x}{2}} \quad \langle 33 \rangle$$

In the case of polyamides similar reactions to (i) and (ii) for esters are used with the amine taking the part of the alcohol e.g.

(iv) $HOOCArCOOH + H_2NArNH_2 + (C_6H_5)_3P + C_2Cl_6$
   $\rightarrow HOOCArCONHArNH_2 + (C_6H_5)_3PO + C_2Cl_4 + 2HCl$

(v) $ClOCArCOCl + H_2NArNH_2 \rightarrow ClOCArCONHArNH_2 + HCl$

Reaction (v) is that often used to prepare lyotropic polyamides such as ⟨44⟩ and gives good yields with high molecular weights.

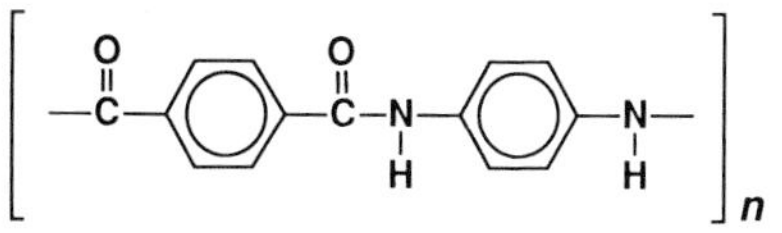

$$\left[-\overset{O}{\underset{\|}{C}}-\hspace{-4pt}\bigcirc\hspace{-4pt}-\overset{O}{\underset{\|}{C}}-\underset{H}{N}-\hspace{-4pt}\bigcirc\hspace{-4pt}-\underset{H}{N}- \right]_n$$

⟨44⟩
poly(*p*-phenyleneterephthalamide)
PPTA

Liquid crystalline *side-chain* polymers are often made by chain-growth reactions such as radical initiated polymerization. Because of their ease of polymerization, preferred monomers are derivatives of acrylic or methacrylic acid, onto which the mesogenic groups are substituted through a flexible spacer. This is shown in the formation of ⟨8ii⟩ from ⟨8i⟩.

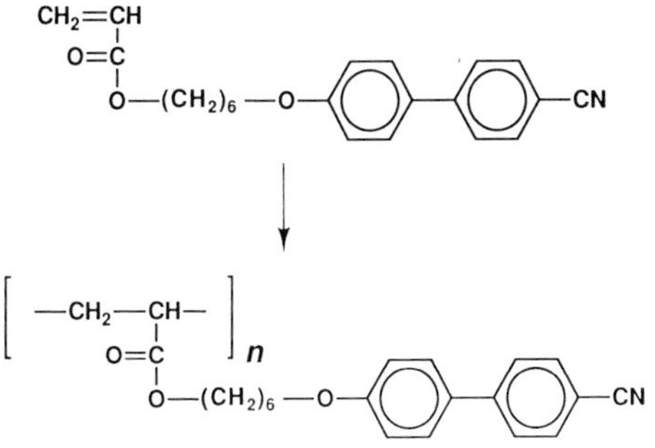

⟨8(i)⟩

⟨8(ii)⟩

The system can be used to make mesogenic copolymers as well as homopolymers, and, by using a certain proportion of monomers substituted at both ends with the acrylic group, it is also possible to cross link the resultant chains.

Synthesis of liquid crystalline polymers in which mesogenic groups are included in both the side chains and the backbone has been achieved by following the normal condensation route. See, for example, Fig. 2.14 in which the mesogenic groups are represented by boxes.

Finally, side-chain mesogenic polymers based on poly(hydrogen methyl siloxane) backbones can be synthesized by grafting the side chains, suitably functionalized with a vinyl group, onto the backbone. The success of the reaction can be compromised if bulky side groups lead to undue steric hindrance, and there can also be difficulties from side reactions (Fig. 2.15).

Radius of gyration (see **Chain dimensions**).

Smectic (see **Friedelian classes**).

Solution thermodynamics The formation of a solution from two components implies that the *free energy* of the solution is less than that of the separate components. The change in free energy as the components are mixed, ΔG_{mix} must be negative for a solution to be stable. It is given by:

$$\Delta G_{mix} = \Delta H_{mix} - T\Delta S_{mix}$$

where ΔH_{mix} is the *enthalpy* change on mixing and is related to the

Fig. 2.14

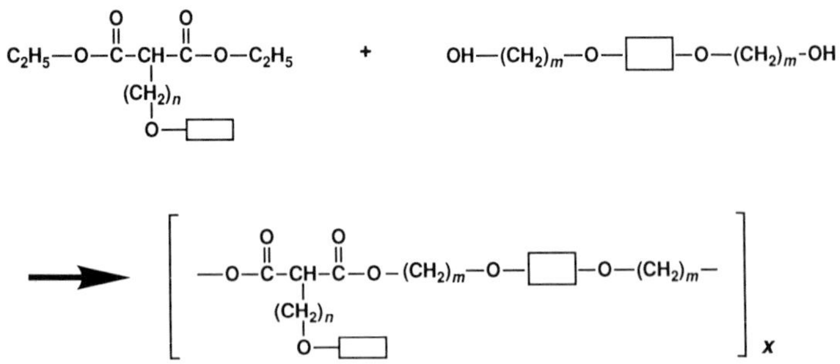

Fig. 2.15

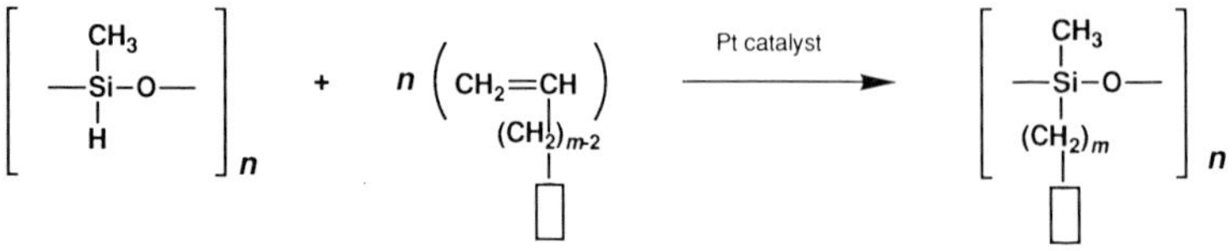

difference between the bonding energy in a mixture of the two different types of molecule and the weighted average of the bonding energy in each of the pure components. In many systems, ΔH_{mix} is positive which means that it opposes solution. The *entropy* of mixing ΔS_{mix} is also positive which means that the term $-T\Delta S$ will, on the other hand, favour solution, especially so at higher temperatures. Lattice model calculations by Flory [2.9] have predicted that the entropy of mixing of a flexible-chain polymer with a small-molecule solvent is given by:

$$\Delta S_{\text{mix}} = -R[n_1 \ln v_1 + n_2 \ln v_2]$$

where n_1 and n_2 are the number of moles of solvent and polymer, and v_1 and v_2 are the respective volume fractions. R is the gas constant.

The enthalpy of mixing changes with the composition of the solution according to:

$$\Delta H_{\text{mix}} = z\Delta\omega_{12}n_1 v_2$$

where z is a coordination number and $\Delta\omega_{12}$ the change in bonding energy on mixing.

Fig. 2.16 A plot of the free energy of mixing, ΔG_{mix}, together with its two component functions, ΔH_{mix} and $-T\Delta S_{\text{mix}}$, as a function of composition for a solvent/polymer system. The double tangent construction defines the compositions of the two component phases. The curves are calculated for a low molecular weight polymer to avoid undue 'skewing' to the solvent end which would make the construction less clear.

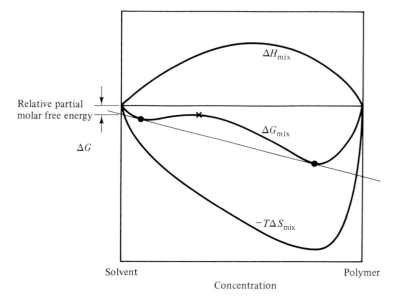

Relative partial molar free energy

ΔG

ΔH_{mix}

ΔG_{mix}

$-T\Delta S_{\text{mix}}$

Solvent   Polymer

Concentration

It is advantageous to recast this relation as:

$$\Delta H_{mix} = \chi R T n_1 v_2$$

χ is known as the *Flory–Huggins parameter*.

Figure 2.16 sketches the dependence of $-T\Delta S$, ΔH and ΔG on the volume fraction of polymer in solution (for positive χ). Note that if ΔH were to be zero or negative, the free energy – composition relation would be convex downwards at all points and thus there would be complete solubility of polymer in the solvent across the full composition range. If however ΔH is positive, as drawn, the free energy curve can have an upward inflection in the mid composition range. A question now arises as to the free energy of a solution of composition in this mid range such as that marked by 'X' on the diagram. However in this case, the free energy would be lower if, instead of a single phase, two solution phases were present each having a composition corresponding, approximately, to one of the minima of the free energy plot. As the two phases would have to be in equilibrium with each other, they would have equal values for a parameter known as the relative partial molar free energy or chemical potential per mole. In graphical terms, this parameter is given, for a particular composition, by the intercept on the vertical axis of the diagram of the tangent to the free energy curve at this composition. The compositions of the two phases in equilibrium are thus given by a *common tangent construction* as shown in the figure. It is clear that a mixture of two phases, each of them a solution of different composition, will have a lower free energy than the single phase solution at 'X'.

Solution thermodynamics form the basis of understanding of lyotropic polymer systems, and are applied in this context in Chapter 4.
Splay (see **Distortions**).
Stereoisomers The repeat units of a homopolymer chain have identical chemical compositions with each atom having the same neighbours. However, it is possible for the units to have left and right handed

Fig. 2.17 Two repeat units of poly(vinylchloride) which are stereoisomers of each other.

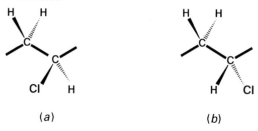

(a)   (b)

versions. Consider the two repeat units drawn in Fig. 2.17. No amount
of local rotation into different conformations will enable the unit on the
left to superimpose onto the one on the right. If a chain is built up from
a sequence of units which are all of one sort or the other then it is said
to be *isotactic*, if alternating, then *syndiotactic*, and if a random mixture
then *atactic*. Regular isomeric sequences preserve the linear periodicity
of the straight molecule and thus permit it to crystallize; by the same
token *atactic* molecules do not crystallize and are the basis of glassy
polymers such as polyvinyl chloride and polystyrene.

Cis-trans isomerism (*cis* = same, *trans* = across) occurs particularly
in polymers which have some double bonds in the backbone about
which rotation is not possible. The two isomers of the repeat unit of
polybutadiene are shown in Fig. 2.18. A chain of all *cis* units makes
natural rubber, one of all *trans*, gutta percha.

Once a unit is connected into a chain, it no longer has the freedom,
that it would have in isolation, to exchange head and tail. Hence if the
polymerization chemistry is independent of head and tail, there are two
ways of connecting identical units into a chain. Furthermore, genuine
stereoisomeric units are possible where the functional groups are
different as could occur with the unit from chloro-hydroxybenzoic acid
(Fig. 2.19).

Strain (see **Mechanical properties**).
Stress (see **Mechanical properties**).

Fig. 2.18 '*Cis*' and '*trans*' repeat units of polybutadiene.

cis   trans

Fig. 2.19 The occurrence of 2,3 isomerism associated with the fact that
there are two non-equivalent substitution sites for the chlorine ('2' and '3'
positions on the benzene ring – or 5 and 6).

(*a*)   (*b*)

Surface orientation The walls of a cell containing a liquid crystal can have a profound influence on the orientation of the director. Two limiting cases of surface-induced texture can be identified. They are illustrated in Fig. 2.20. In the *homogeneous* or *planar* texture the molecules lie parallel with the interface, whereas in the *homeotropic* texture the molecules are aligned normal to the surface. Homeotropic alignment is favoured when the end groups of the molecules have a particular chemical affinity for the cell walls which are usually glass. T_g (see **Glass transition**).

Thermotropic liquid crystals The term 'thermotropic' describes

Fig. 2.20 (*a*) Homogeneous (planar) and (*b*) homeotropic textures induced by specific surface orientation effects.

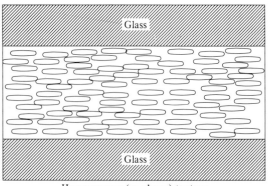

Glass

Glass

Homogeneous (or planar) texture

(*a*)

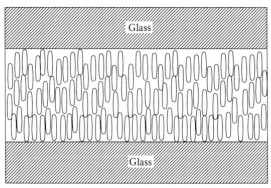

Glass

Glass

Homeotropic texture

(*b*)

substances which show liquid crystallinity in a particular temperature range, and without the need for the addition of solvent molecules. The term is particularly used to distinguish liquid crystalline phases from those, known as *lyotropic*, which contain solvents. By far the majority of small-molecule liquid crystals studied are thermotropic, but both types have very important applications in liquid crystalline polymers.
Twist (see **Distortions**).

Viscosity Viscosity, η, is a measure of the resistance of a fluid to shear flow. It is defined as the shear stress necessary to produce a unit shear strain *rate*, i.e.

$$\eta = \sigma_{xy}/(\partial \varepsilon_{xy}/\partial t)$$

The units of viscosity are thus $N \, s \, m^{-2}$, also written as Pa s. An alternative (non-SI) unit of the poise (P) is also sometimes used, with $1 \, P = 0.1 \, N \, s \, m^{-2}$.

Where the viscosity is independent of the strain rate, the flow is said to be *Newtonian*. Many polymeric materials show strongly non-Newtonian flow properties, and such behaviour is also a characteristic of liquid crystalline polymers. Very often the material is found to be *shear-thinning*, so that the viscosity decreases as the shear rate is raised. This has important implications for processing.

For liquid crystals, because the liquid is not isotropic, a single viscosity coefficient is not sufficient. In this case, several different parameters may be chosen to describe the situation, but the most commonly used are the three *Miesowicz viscosities*.

3 Stability of liquid crystalline polymers

3.1 Introduction

A molecule is said to be *mesogenic* if it is able to form liquid crystalline phases. However, the actual occurrence of liquid crystallinity depends on other factors as well. In *thermotropic* systems, the liquid crystalline phase only exists within a particular temperature range. It lies between the crystal melting point, T_m (or in cases where crystallinity is absent, the glass transition temperature, T_g), and the so-called upper transition temperature where the liquid crystalline phase reverts to an isotropic liquid, $T_{lc \to i}$.

This chapter focuses on the relationship between the chemical structure of mesogenic polymer molecules and the temperature range over which they can form liquid crystalline phases. It does not deal with the different classes of mesophase: nematic, smectic etc., which can occur in polymeric materials. They are fully described in Chapter 5.

It is helpful to view liquid crystalline polymers as being constructed from the same rigid mesogenic groups as form small-molecule liquid crystals. Figure 3.1 (overleaf) illustrates some of the wide range of molecular architectures which can be created through the combination of mesogenic groups and other types of sequence within a polymeric system. The most basic construction is simply to link the rigid mesogenic groups end to end to form a rigid polymer chain. However, in practice this straightforward approach leads to a significant problem; for, while T_m and $T_{lc \to i}$ are both increased as a result of the polymerization and the *range* of liquid crystal stability is in principle maintained or even enhanced, thermal degradation of the polymer tends to curtail this range and often prevents the mesophase being observed at all.

The reduction of the transition temperatures into a useful working range without destroying the mesophase stability completely is one of the primary objectives in the design of liquid crystal polymers. There are various approaches to the problem, and in thermotropic systems these all involve aspects of molecular design. One route is to make the chain less stiff by a controlled amount through introducing rather more mobile linking groups between the rigid units. Alternatively, different units can be randomly positioned along the chain, and if these added units are themselves mesogenic one gains the particular advantage of reducing the crystal melting point without any associated reduction in the stability of the mesophase. A further variant involves the connection of the mesogenic units through spacers consisting of lengths of flexible polymer molecule. Such *flexible spacers* may be incorporated between the mesogenic groups to give a single linear molecule; or a flexible polymer chain can be used as a backbone to which the mesogenic sections are added as side groups, the connections themselves often being made through short sequences of flexible links. There are of course several other variants to this theme. Such systems, both main-chain and side-chain, can give electro-optical properties which provide useful modifications to those seen in small-molecule materials.

Another way is to proceed via a solution route. A liquid crystalline system in which transition temperatures are brought down by the addition of a low molecular weight solvent is known as *lyotropic*, and this term is also applied to polymeric systems. Lyotropic (i.e. liquid crystalline) solutions enable very rigid molecules to be handled, although wet processes have their own particular inconveniences.

We shall now consider these various approaches in more detail, and include specific examples.

3.2   Factors limiting liquid crystallinity

3.2.1   *Crystal melting point*

In general, the reduced freedom a monomer unit has to move once it is a part of a chain leads to a pronounced increase in melting temperature; for example, ethylene melts at $-160\ ^\circ\text{C}$ whereas linear polyethylene as normally crystallized melts at $130\ ^\circ\text{C}$. At the melting point, T_m, the crystal phase is in equilibrium with the melt and thus it is possible to write:

$$T_m = \Delta H / \Delta S \tag{3.1}$$

where ΔH is the difference in enthalpy between the melt and the crystal, the presence of 'better' intermolecular bonds leading to the lower value in the crystal. ΔS is the corresponding entropy difference and reflects the greater randomness and freedom of motion in the melt.

The increase in entropy when ethylene melts is $32.2\ \text{J mol}^{-1}\ \text{K}^{-1}$ and is associated with the freedom of the molecule to change position and to rotate. Substitution of the melting point (in degrees K) into equation (3.1) gives an enthalpy change ΔH of $3.35\ \text{kJ mol}$. If the enthalpy change is measured for a polymer molecule it is surprisingly similar to that for the monomer as long as it is expressed *per mole of monomer units* and not per mole of polymer molecules. For example ΔH per

Fig. 3.1 Various types of liquid crystalline polymer molecule.

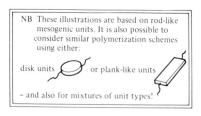

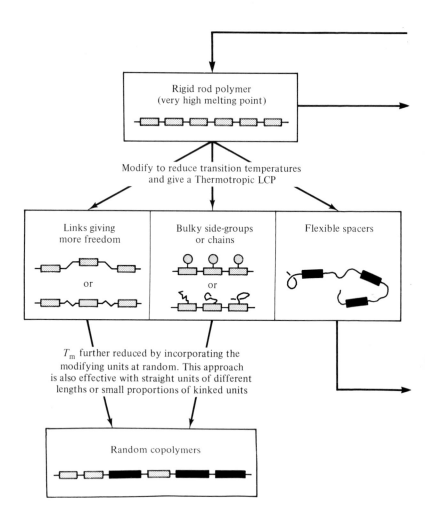

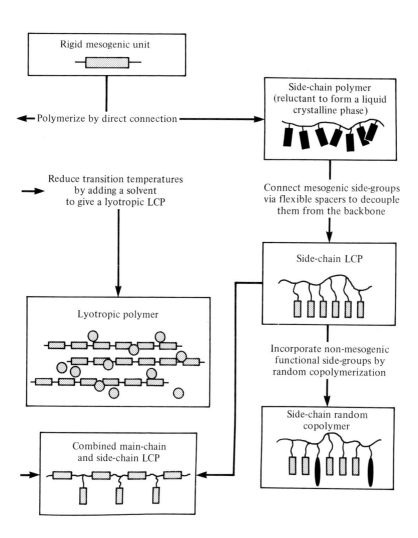

Table 3.1 *Thermodynamic data for a rod-like molecule* [3.1]

<9>

Transition	T_m(°C)	ΔS J/mol/K	ΔH kJ/mol
crystal–liquid crystal	250	21.1	11.1
liquid crystal–isotropic	392	0.72	0.48

monomeric unit for polyethylene is 3.29 kJ mol^{-1} monomer. On the other hand ΔS remains approximately independent of molecular size, if the molecule is assumed to be rigid. Thus ΔS/monomeric unit should be expected to decrease in inverse ratio to the number of units in a chain. Consideration of equation (3.1) will immediately suggest that the melting temperature of any high polymer will be emphatically 'off-scale' if the long molecule is considered to be a completely rigid rod – *which is the key point*. Polyethylene is of course a flexible molecule and the change in entropy on melting of $\Delta S = 7.9$ J mol^{-1} of monomer K^{-1} represents almost entirely the conformational freedom in the melt, with essentially negligible contributions from rotation or translation of the molecule as a whole. Any reduction in flexibility can be expected to lead to significant increases in melting temperature. One has only to look at poly(ethyleneterephthalate) ($T_m = 267$ °C) and poly(tetrafluorethylene) ($T_m = 327$ °C) to see the consequences of increasing molecular stiffness.

For polymer molecules which are sufficiently rigid to be candidates for liquid crystalline behaviour, the implication is clear: their melting points are likely to be high and possibly in excess of the temperature at which the polymer begins to decompose. The fact that the liquid phase in equilibrium with the crystal at the melting temperature may be a mesophase with long range orientational order is likely to have relatively little influence on ΔS and ΔH for crystal melting. The entropy and enthalpy changes at the liquid crystalline to isotropic transition in the liquid are much less than at crystal melting. For example, the small molecule $\langle 9 \rangle$ has values of ΔS and ΔH which are about twenty times smaller at the liquid crystalline to isotropic transition than at the crystal melting point (Table 3.1).

The rapid increase in melting point with length can be illustrated by considering a series of molecules of similar structure and increasing chain lengths as depicted in Table 3.2. Molecule $\langle 10 \rangle$ has a crystal melting point of 132 °C and melts to give an isotropic phase without any intervening mesophase. The addition of a further phenyl ester

Table 3.2 *Effect of length of rod-like molecules on transition temperatures* [3.2]

Compound	T_m(°C)	$T_{lc \to i}$
⟨10⟩	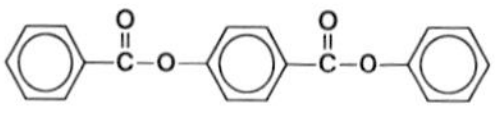 132	–
⟨11⟩	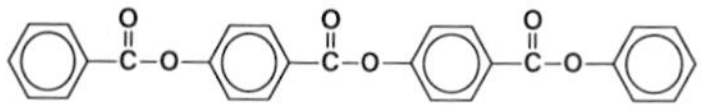 176	251
⟨12⟩	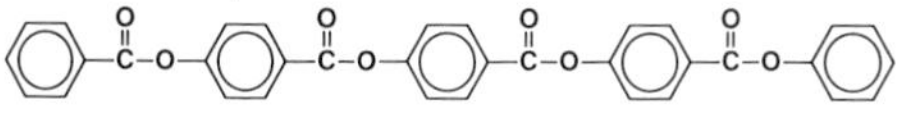 220	391

group to form the four-ring derivative ⟨11⟩ increases the melting point to 176 °C and also leads to liquid crystalline stability which persists for a further 75 °C above the melting point. The five-ring compound ⟨12⟩ still continues the upward trend of transition temperatures with a melting point of 220 °C and the liquid crystallinity persisting for a further 170 °C up to 391 °C.

With the five-ring 'chain' a significant proportion of the entropy change on melting will be due to the onset of chain motion within the liquid crystalline phase, in particular the rotations of the *para* (*p-*) linked phenylene groups and the crankshaft motion of the esters. This contribution to ΔS will be per repeat unit, not per molecule, and will not decrease (for a given mass of material) with increasing chain length. There will thus be a lower limit to ΔS which will be the effective value in high polymers with 100 or more units in the chain. The trend of increasing melting point with increasing length will therefore not continue indefinitely, and the melting point of a molecule such as poly(hydroxybenzoic acid) (PHBA) ⟨13⟩ need not be exceptionally high. In fact the exact melting point of PHBA is a matter for debate. X-ray studies show a retention of order up to 480 °C [3.3], although there is a major transition at 350 °C to a pseudo-hexagonal crystal phase within which there is a considerable level of chain motion. Recent DSC measurements [3.4] indicate that there is a high temperature transition at 445 °C although it is not clear whether a liquid crystalline phase is stable at this temperature. However, while one may have to assume the melting point of PHBA to be around 500 °C, a temperature at which thermal degradation interferes with precise measurement, there is no doubt that the closely related

$$\left[-O-\!\!\bigcirc\!\!-\overset{\overset{\displaystyle O}{\|}}{C}- \right]_n$$ ⟨13⟩   PHBA

poly(hydroxybenzoic acid)

$$\left[-O-\!\!\bigcirc\!\!\bigcirc\!\!-\overset{\overset{\displaystyle O}{\|}}{C}- \right]_n$$ ⟨14⟩   PHNA

poly(hydroxynaphthoic acid)

polymer, poly(hydroxynaphthoic acid) (PHNA) ⟨14⟩, melts to a smectic liquid crystalline phase at 440 °C [3.5]. The lower melting point of the naphthoic polymer being ascribable to the extra crankshaft type motion associated with the rotation of the 'side-step' in the 1,6 linked naphthoic ring.

In broad terms, the direct (*para*) linkage of aromatic groups to form a linear homopolymer molecule, such as poly(*p*-phenylene) ⟨15⟩, will

$$\left[-\!\!\bigcirc\!\!- \right]_n$$ ⟨15⟩   PPP

poly(*p* - phenylene)

give a crystal structure which will not melt below its decomposition temperature. Linkage through ester groups means that additional chain motion is possible in the liquid crystalline phase with the result that the melting point is reduced, albeit not sufficiently to permit melt processibility without significant risk of thermal degradation. Other molecular modifications are necessary to bring the melting point of these types of molecule into the optimum range and these are discussed in detail in Section 3.3 below. Firstly, however, it is useful to look at the effect on the liquid crystal → isotropic transition temperature ($T_{lc \to i}$) of simply connecting mesogenic groups together to form linear polymer chains, for a feature of molecular design is of little value if it reduces ($T_{lc \to i}$) more effectively than the crystal melting temperature and thereby eliminates the liquid crystalline phase altogether.

3.2.2 *Liquid crystalline → isotropic transition temperature*

Whereas the increase in melting point as more and more mesogenic units are added together is essentially a function of the size of the molecule, the temperature at which the liquid crystalline phase (if there is one) transforms to the isotropic liquid, $T_{lc \to i}$, depends mainly on the anisotropy of molecular shape. Increasing the length of a rigid chain molecule increases both its size and axial ratio. Looking again at molecules ⟨10⟩ and ⟨11⟩, not only is the melting point increased, but the greater anisotropy of the four-ringed molecule establishes mesogenic properties giving a $T_{lc \to i}$ of 251 °C. Correspondingly the $T_{lc \to i}$ for the five-ringed molecule $\langle 12 \rangle$ is 391 °C. While the relative increase in the two transition temperatures with increasing length is not easy to predict with any precision and depends on the exact balance between ΔS and ΔH, it is generally true that the upper transition increases more rapidly than the crystal melting point.

Having established the trend in $T_{lc \to i}$ with increasing molecular length, it is instructive to consider the case of very long and completely rigid rods. In the isotropic melt with its absence of long range orientational order (liquid crystallinity), their 'packing' must be reminiscent of the proverbial *bag of nails*. Steric problems will lead to very low packing density and thus a comparatively large ΔH, and the relatively small change in ΔS associated with the onset of orientational freedom, when the molecule is both large and rigid, will combine to suggest that $T_{lc \to i}$ values will be very high indeed; in fact, as with predictions of T_m for similar rigid molecules, they will be off-scale.

Another view of the influence of rigid chain length emerges from statistical models of rigid rods. Anticipating the theoretical treatments outlined in the next chapter, it is worth noting that Flory and Ronca [3.6] have demonstrated for an undiluted system, i.e. one which is thermotropic rather than lyotropic, that the ordered liquid crystalline phase will always be stable with respect to the isotropic if the axial ratio, x (length/diameter) of the rods exceeds 6.4. This ratio is arrived at by considering only the repulsive, shape dominated, forces between the molecules. Incorporation of possible directional attractive forces, such as might arise from dipolar groups, leads to the prediction that rods of lower axial ratio can form mesophases, although this tendency will be opposed with increasing temperature.

Polymer molecules will have axial ratios far in excess of 6.4 if they are straight and rigid. The question is, how much is the *effective* axial ratio reduced by increasing chain flexibility? It seems that the effective rigid length of a long molecule can be described fairly well by the parameter known as the *persistence length*, q. It is normally measured using the methods of either light scattering or small angle X-ray scattering with the polymer in dilute solution although, in the case of rigid polymers it is often difficult to find suitable solvents. There is also the additional problem that the presence of solvent may change the preferred conformation of the molecule and thus its rigidity. A selection of measured persistence lengths, q, together with estimates of the effective persistence ratios $x_p (= q/d$ where d is molecular diameter), is listed in Table 3.3.

If we are to relate persistence ratio of a chain to the axial ratio of a

Table 3.3 *Persistence length, q, and persistence ratio, x_p, of a selection of high molecular weight polymers*

Polymer		q(nm)	$x_p = q/d$
<16> polyisoprene	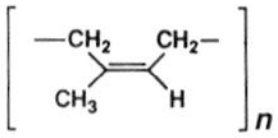 $\left[\begin{array}{c} -CH_2 \quad CH_2- \\ CH_3 \qquad H \end{array}\right]_n$	0.76	1.5
<17> poly(vinylacetate)	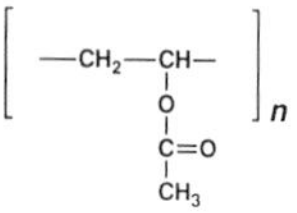 $\left[\begin{array}{c} -CH_2-CH- \\ O \\ C=O \\ CH_3 \end{array}\right]_n$	0.95	1.4
<18> poly(vinylbromide)	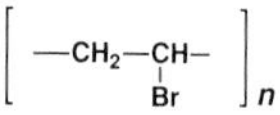 $\left[\begin{array}{c} -CH_2-CH- \\ Br \end{array}\right]_n$	1.25	2.2
<19> poly(tetrafluorethylene) PTFE	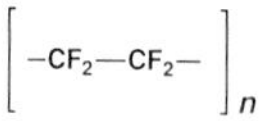 $\left[-CF_2-CF_2-\right]_n$	2.23	4.5
<20> cellulose nitrate	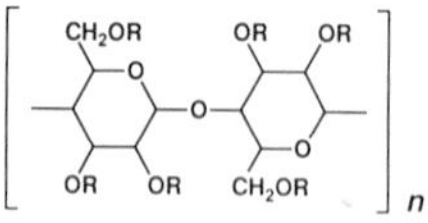 [cellulose ring structure with CH$_2$OR, OR, OR groups] R is NO$_2$	2.2	4
<21> hydroxypropyl cellulose HPC	[cellulose ring structure with CH$_2$OR, OR, OR groups] R is $-CH_2CHCH_3$, OH	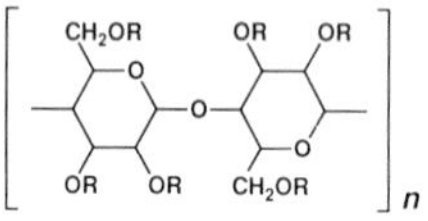 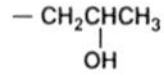 8.5	19
<22> poly(γ-benzyl-L-glutamate) PBLG	$\left[\begin{array}{c} O \; H \; H \\ -C{-}N- \\ R \end{array}\right]_n$ R is $-(CH_2)_2-\overset{O}{\overset{\|}{C}}-O-CH_2-$ (benzene ring)	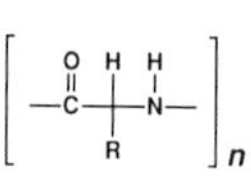 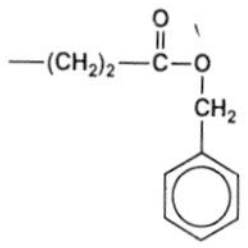 20.0	16
<23> poly(p-benzamide) PBA	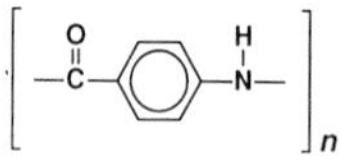 $\left[\begin{array}{c} O \qquad\qquad H \\ -C-(\text{benzene ring})-N- \end{array}\right]_n$	40.0	80

n.b. the substitution of R for H in the celluloses may not have been complete.

small rod-like molecule and thus assume that the limiting axial ratio for liquid crystallinity for rods of 6.4 is relevant to polymer chains, then we might expect that the final three polymers in Table 3.3 could be liquid crystalline above ambient temperature and possibly above their melting points if thermal decomposition does not intervene. This is indeed the case with hydroxypropyl cellulose (HPC) and poly(γ-benzyl-L-glutamate) (PBLG), although the crystallinity in poly(p-benzamide)

(PBA) persists up to the decomposition temperature masking the liquid crystallinity. Cellulose nitrate and poly(tetrafluorethylene) (PTFE) are border-line with respect to the critical persistence ratio and could possibly show liquid crystalline order. However, the first decomposes emphatically at 225 °C, and there is, as yet, no compelling evidence for its existence in the latter.

It might be argued that liquid crystalline polymers in which the mesophase–isotropic transition temperature is infinite, or at least experimentally inaccessible, should not be classified as thermotropic [3.7]. However, the existence of a glassy or a crystalline state which melts to give a liquid crystalline mesophase is, not unreasonably, taken as sufficient justification for the thermotropic label. We have adopted this latter view, if only because 'thermotropic' is now common parlance for many practical liquid crystalline systems which nevertheless decompose before any transition to the isotropic phase can be detected.

3.3 Chain modification to control thermotropic liquid crystalline polymer stability

3.3.1 *Introduction*

The purpose of this section is to illustrate the principles discussed above with specific examples of chain modification, and to show how polymers can be designed with liquid crystalline stability in an optimum temperature range. Just what this range is depends on the end use in view. For example, a mouldable liquid crystalline polymer for a high temperature structural application would ideally have a melting point between 350 °C and 400 °C, the upper temperature being determined largely by the limitations of conventional processing machinery. It would have a liquid crystalline melt, although a comparatively narrow temperature range between T_m and the upper transition to the isotropic state is acceptable, while good resistance to thermal degradation at these temperatures is an equally important attribute. For a polymer which is to be used at ambient temperatures, perhaps for a structural application like sheathing optical fibres, or as the active element of a non-linear optical device, a lower processing temperature is possible and the polymer would be designed to have a melting point in the region of 200 °C, with a mesophase stability range covering at least 50 °C above this temperature.

3.3.2 *Types of modification*

Recalling equation (3.1), the melting temperature T_m will be reduced if the entropy change on melting ΔS can be increased, or the enthalpy change ΔH decreased, or both. A larger ΔS will be achieved if the conformational freedom within the mesophase can be increased, for example by introducing short links with at least some flexibility into the chain. An effective way of decreasing ΔH is to insert modifying units at random positions along the chain, thus forming a random copolymer in which crystallinity is frustrated. Often both principles are active at once. Where the modifying units are not themselves mesogenic, the mesophase–isotropic transition temperature will be reduced along with the melting point. Indeed, it is usually brought down more rapidly, so that, as the proportion of the non-mesogenic unit is increased, the range of liquid crystalline stability is compressed and eventually eliminated.

There is virtually no limit to the number of ways polymer chains can be modified through the incorporation of different chemical units in different positions. However it is helpful, in considering the influence of molecular architecture on liquid crystallinity, to divide the possible types of chain modification into three groups:

(a)   addition of modifying units at regular intervals along the mesogenic chain

(b)   addition of modifying units at random positions along the chain to give a random copolymer

(c)   addition of sequences of non-mesogenic units to give similar lengths of flexible chain between the mesogenic units.

3.3.3   Polyesters as an example

Aromatic polyesters are the most important class of thermotropic liquid crystalline polymers developed for structural applications, and they serve as a good example of the ways in which chain modification can be used to control mesophase stability. The regular molecules used as a basis for looking at modification are poly(hydroxybenzoic acid) (PHBA) ⟨13⟩, and its close analogue, poly(p-phenyleneterephthalate) (PPT) ⟨24⟩ in which the ester groups alternately face left and right along the chain. The melting points of these two homopolymers are in the temperature range where thermal decomposition is setting in, being around 500 °C and possibly as high as 600 °C [3.8]. Also, in each case the liquid crystalline to isotropic transition is well above the decomposition temperature.

The ester link is drawn in more detail in Fig. 3.2. The virtual bonds about which each of the phenylene units can rotate are not collinear, each ester link giving a sideways displacement of about 1.34 Å. Rotation about the single bond joining the oxygen atom to the carbonyl group (C=O) is difficult as it takes on some of the character of the adjacent double bond due to resonance. The ester linkage thus provides little opportunity for the chain to kink. In fact it is often used in small-molecule liquid crystals, where its tendency to promote mesogenicity is even ranked above that of a single bond directly linking the adjacent phenyl groups [3.9]. Conformational energy calculations (e.g. [3.10]) indicate that the phenylene group, which is attached to the ester group through the singly bonded oxygen, is prevented from rotating into the same plane as the ester by steric interaction between one of the *ortho* hydrogens on the phenyl ring and the doubly bonded oxygen atom. However, the side step associated with the ester provides an additional measure of conformational freedom in the liquid crystalline phase through crankshaft type motions, in that rotation about bonds **R** or **T** (Fig. 3.2), will cause the second phenyl group to move laterally with respect to the first.

A selection of modifying units for the aromatic ester chains is shown in Table 3.4. Some are rigid and straight, for example biphenyl, others introduce side steps, kinks, bulky side groups or flexibility. It is important to appreciate that their effectiveness in controlling the thermotropic transition temperatures depends as much on the way in which they are positioned within the polymer chain, as on their own intrinsic properties.

Fig. 3.2 The geometry of an ester link between two phenyl groups. The length dimensions are in ångströms.

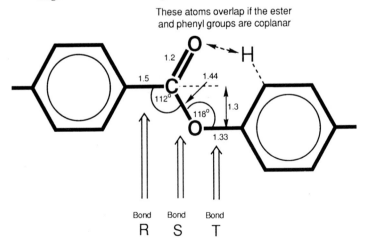

These atoms overlap if the ester
and phenyl groups are coplanar

Bond   Bond   Bond
 R   S   T

Table 3.4 *Examples of modifying units*

	Unit	Type	Modifying effect
(i)		basic phenyl unit	
(ii)		biphenyl	length variation
(iii)		naphthoic	length and side step
(iv)	R	ring substitution	poorer crystal packing
(v)	$-(CH_2)_2-$	ethylene link	length and flexibility
(vi)		*meta* link	kink
(vii)	CH_2	single tetrahedral bond	kink

3.3.4 *Modification with regularly positioned units*

If a chain is built with benzoic units regularly alternating with another mesogenic unit such as hydroxynaphthoic or biphenyl, then there will be comparatively little effect on the transition temperatures. The added side step associated with the naphthoic unit will reduce the melting point a little (the melting point of its homopolymer, PHNA, is 440 °C), while the effect of biphenyl is likely to be slightly in the other direction. It is clear that the modifying units which are able to reduce the transition temperatures significantly when introduced regularly along the chain are either the flexible, non-mesogenic ones, or ones with substantial side groups.

One of the most useful of short flexible links is $-(CH_2)_2-$, which in its *trans* conformation does not bend the molecule. It is effective in reducing T_m by increasing the conformational entropy in the liquid crystalline phase, but it also reduces $T_{lc \to i}$ through a drastic decrease in rigidity and thus effective persistence ratio of the chain. Its influence in this respect is best illustrated by comparing the melting points of poly(p-phenyleneterephthalate) $\langle 24 \rangle$ with, firstly, poly(ethyleneterephthalate) (PET) $\langle 25 \rangle$ which has one $-(CH_2)_2-$ link per phenylene

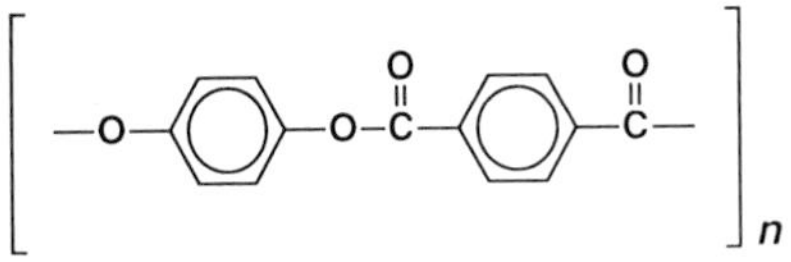

<24>

poly(*p*-phenyleneterephthalate)
PPT

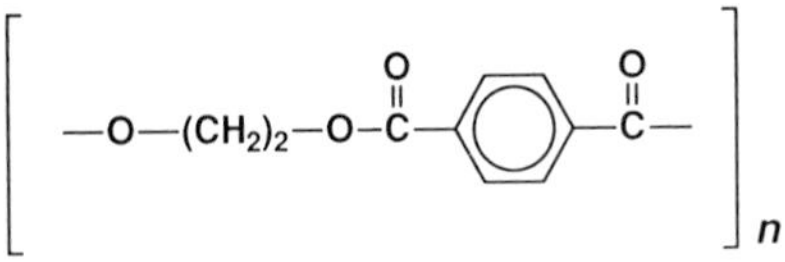

<25>

poly(ethyleneterephthalate)
PET

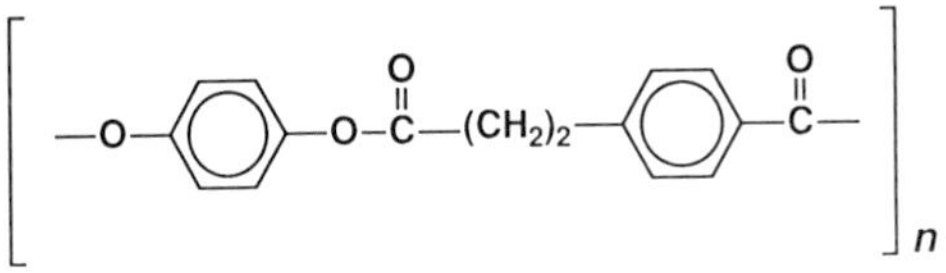

<26>

group and shows no liquid crystalline phase, and secondly, with poly(4-carboxybenzene-propionic acid) ⟨26⟩ which has two phenylenes per —$(CH_2)_2$— link. The presence of one —$(CH_2)_2$— link for two phenylenes reduces the melting point from above 500 °C to 425 °C but does not bring $T_{lc \to i}$ down into this range. The polymer thus has a thermally accessible liquid crystalline phase. On the other hand, one —$(CH_2)_2$— per phenylene unit as in PET reduces the melting point to 265 °C but also effectively brings down $T_{lc \to i}$ to an even lower temperature and thus prevents liquid crystallinity.

The incorporation of kinked units such as *meta* linked phenyl rings also has a drastic influence on the stability of the liquid crystalline phase, although its effect on T_m is not as pronounced as in the case of the flexible —$(CH_2)_2$— unit. For example, an alternating polyester of *para* and *meso* linked rings ⟨27⟩ is not liquid crystalline although its melting point is of the order of 380 °C.

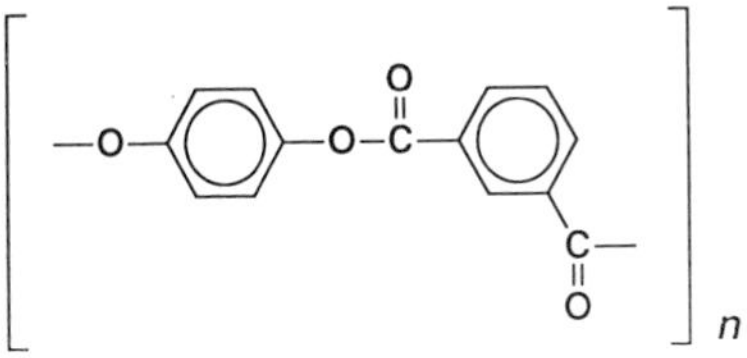

<27>

Substantial reductions can also be achieved by alternating the basic mesogenic units with similar units which are singly substituted on the *p*-phenylene ring. The bulky side group influences the transition temperatures in several ways. It effectively increases the chain diameter leading to a modest decrease in both the persistence ratio and the upper transition temperature. The localised 'thickening' of the chain at the side group position has a general tendency to reduce the efficiency of chain packing within the crystal lattice and hence the melting point. However, this effect is much enhanced by the fact that the side group is substituted either at the 2 (or 6) position or the 3 (or 5) position of the ring, which means that a degree of randomness can be introduced into the chain. In the case of di-acid or di-ol units, the absence of 'left–right' symmetry of the substituted ring opens up the possibility of head-to-head, head-to-tail isomerism. Without stereospecific control the molecule will polymerize as an atactic chain which will further inhibit crystallization. In molecules $\langle 28 \rangle$ and $\langle 29 \rangle$, the melting points have been reduced to 370 °C and 340 °C respectively. The random positioning of the larger phenylene group has the greater influence, and the polymer has a melting point which is low enough to provide access to the liquid crystalline melt at more convenient processing temperatures.

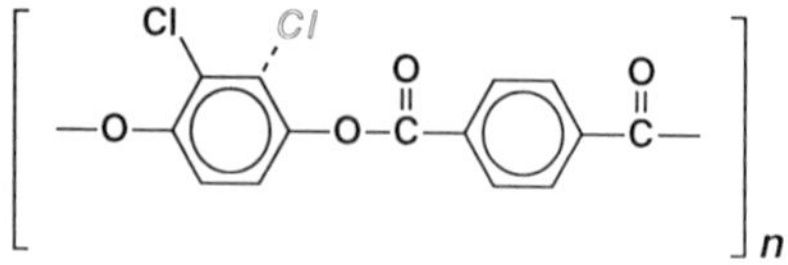

$\langle 28 \rangle$

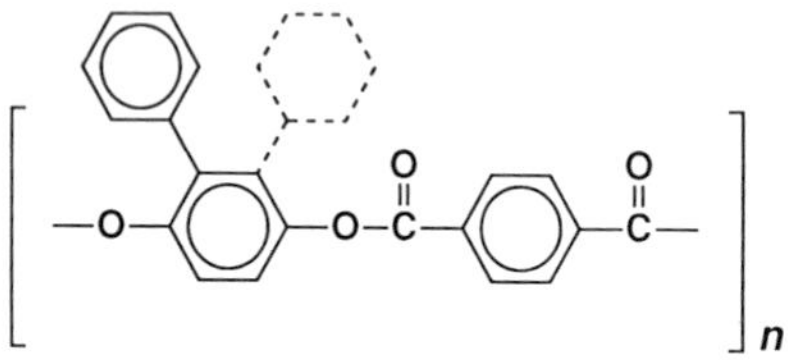

$\langle 29 \rangle$

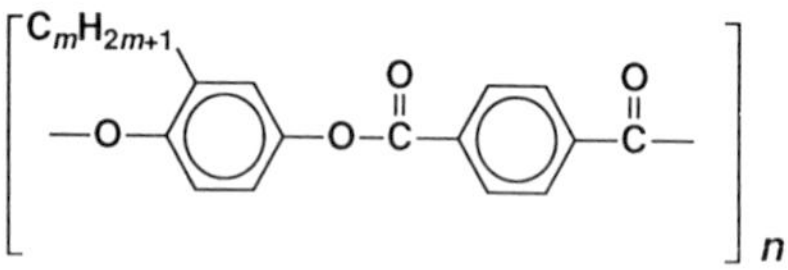

$\langle 30 \rangle$

The next step in the substitution story is to consider the addition of non-mesogenic chains to straight, rigid backbones, to produce the so called 'hairy rod' molecules. The effect on T_m and $T_{lc \to i}$ of adding different lengths of alkyl side chains, $-(CH_2)_m-$, onto one ring of poly-(phenyleneterephthalate) $\langle 30 \rangle$ is shown in Fig. 3.3. Perhaps not surprisingly, longer side chains give lower transition temperatures, leading for $m > 16$ to layer structures, where a separate low temperature transition can be identified with the melting of segregated side-chain material, and eventual loss of liquid crystallinity [3.11]. Similar observations have been made for related polymers with up to four substituted chains per backbone repeat, and for amide linked backbones, although in the latter case greater side-chain lengths were necessary to reduce the crystalline melting point to below the decomposition temperature. In some respects it is possible to view the side chains as a sort of *bound solvent*. The lowering of transition temperatures is thus due to a reduction in the volume fraction and hence anisotropic interaction of the rigid chains. The analogy with solvent based lyotropic systems (Section 3.4) is intriguing, although the molecular attachment of the diluting component means that it does not have to be a solvent for the rigid chain in the thermodynamic sense. For example, polyethylene and poly(phenyleneterephthalate) are immiscible. Furthermore, there would not be the requirement of solvent removal to form a structurally useful material, although it should be borne in mind that the presence of the side chains will also tend to

Fig. 3.3 Plot showing the influence of the length of the substituted side chain of molecule ⟨30⟩ on the stability range of the liquid crystalline phase. The breadth of the melting band is associated with the observation of double melting peaks in DSC traces of some of the polymers. (The diagram is based on data from reference [3.11].)

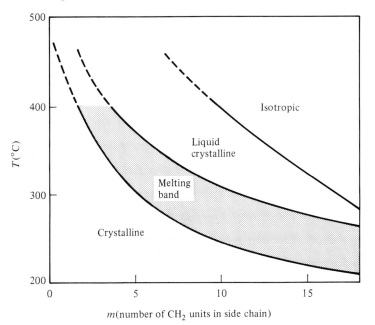

m(number of CH_2 units in side chain)

'dilute' the high levels of axial strength and rigidity associated with the rigid chains (Chapter 8).

3.3.5   *Modification by random copolymerization*

The effectiveness of units in reducing the crystalline melting point is greatly enhanced if the chain is modified by units incorporated in random positions along its length. The destruction of chain periodicity inhibits crystallization, markedly reducing both crystallinity and melting point, without necessarily leading to any additional loss of mesogenicity and reduction in upper transition temperature. Take for example the random copolymer ⟨31⟩, of hydroxybenzoic acid (HBA) with equal proportions of terephthalic acid and p, p'-biphenol. Its melting point with a mole fraction of HBA units (x) of 0.33 is 380 °C [3.12]. Thus the non-regular insertion of 25% biphenyl units into the molecule achieves a reduction in T_m of around 150 °C. The biphenyl is rigid and produces neither side step nor kink. It gives a reduction in T_m because it is longer than the phenyl units, and the long range periodicity and hence crystallizability of the molecule is impaired. Modifying units, other than those involving substitution on the phenylene, can normally be expected to differ in length from the basic phenylene unit, so the 'length effect' will usually be a major factor in determining the reduction in melting point on random copolymerization. Systems in which the benzoic ester molecules are modified by biphenyl units form the basis of the *Xydar* series [3.13] of mouldable liquid crystalline polymers.

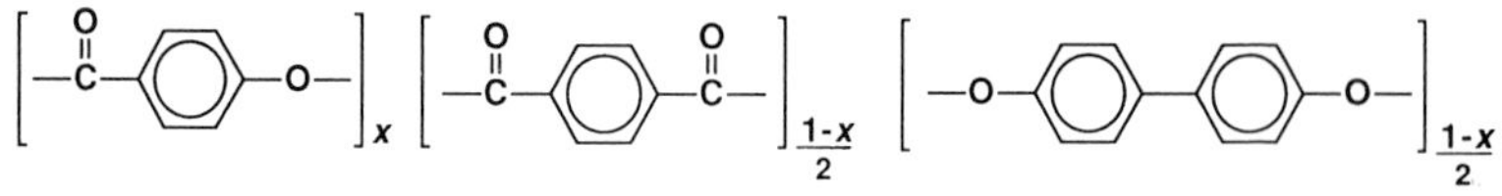

⟨31⟩

Another important thermotropic system is based on random copolymers of hydroxybenzoic acid and hydroxynaphthoic acid (HNA) ⟨32⟩. It was first commercialized in 1985 as the *Vectra* series of mouldable polymers [3.14]. The primary influence of randomly positioned naphthoic units in reducing T_m again results from their greater length than the benzoic units, although the side step associated

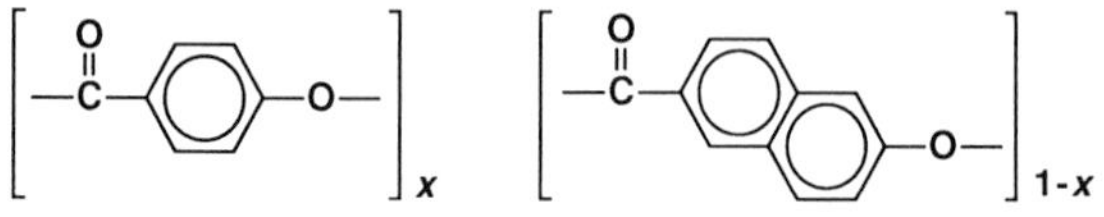

⟨32⟩

with the unit will also contribute. Figure 3.4 is a plot of melting point against molar composition of this random copolymer. The minimum melting point, at 42 mole % hydroxynaphthoic acid units, is 252 °C, and the system is liquid crystalline for all compositions. Naphthalene based units prove very efficient modifiers, not only because they can reduce T_m so effectively but also because they do not detract from the excellent mechanical properties and thermal and chemical resistance germane to the poly(hydroxybenzoic acid) molecule. Their only drawback is that they are considerably more expensive than the benzoic ones.

A particularly good example of the efficiency of random copolymerization in reducing melting point is provided by the system based on *p*-hydroxybenzoic acid and equal amounts of terephthalic acid and ethylene glycol ⟨33⟩. A phase diagram of temperature against composition is shown in Fig. 3.5. The homopolymer based on terephthalic acid and ethylene glycol at the left of the diagram is PET, which is not liquid crystalline. The effect of increasing the proportion of the rigid, linear aromatic groups is to decrease rather than increase the melting point, as the tendency for the melting point to increase with increasing chain rigidity is more than compensated by the crystalline

Fig. 3.4 The variation in melting point with composition of a thermotropic random copolymer ⟨32⟩ based on hydroxybenzoic acid (HBA) and hydroxynaphthoic acid (HNA) [3.15].

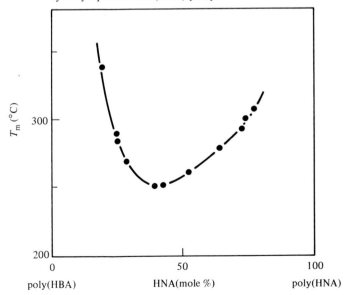

disruption stemming from the random copolymerization. Between 30 and 40 mole % HBA the system becomes thermotropic and $T_{lc \to i}$ increases rapidly with HBA content. The minimum melting temperature of 190 °C occurs at around 60 mole % HBA. At higher concentrations, T_m moves up very rapidly indeed and the system rapidly loses its liquid crystalline processibility.

$$\left[-\overset{\overset{\displaystyle O}{\|}}{C}-\bigcirc-O- \right]_x \left[-\overset{\overset{\displaystyle O}{\|}}{C}-\bigcirc-\overset{\overset{\displaystyle O}{\|}}{C}- \right]_{\frac{1-x}{2}} \left[-O-(CH_2)_2-O- \right]_{\frac{1-x}{2}} \qquad \langle 33 \rangle$$

The introduction of kinked units by random copolymerization is an especially effective means of reducing the melting point. Such units also oppose the mesogenicity of the melt in reducing the persistence length and effective axial ratio. Polyesters based on *p*-hydroxybenzoic acid modified by random copolymerization with isophthalic acid and hydroquinone $\langle 34 \rangle$, the *meta* linkage of the isophthalic acid providing the kink, are an important example of this type of modification. The

Fig. 3.5 Schematic phase diagram showing the stabilization of a liquid crystalline phase and the initial reduction in melting point as hydroxybenzoic acid units are copolymerized with equal amounts of terephthalic acid and ethylene glycol to give the random copolymer $\langle 33 \rangle$. (After [3.16] with other unpublished data.)

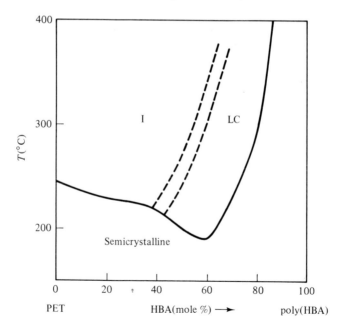

$$\left[-C \bigcirc O- \right]_x \left[-C \bigcirc {}^{C-} \right]_{\frac{1-x}{2}} \left[-O \bigcirc O- \right]_{\frac{1-x}{2}}$$

⟨34⟩

$$\left[-C \bigcirc O- \right]_x \left[-C \bigcirc C- \right]_{\frac{1-x}{2}} \left[O \bigcirc {}^{CH_3}_{C}{}_{CH_3} \bigcirc O \right]_{\frac{1-x}{2}}$$

⟨35⟩

phase diagram, Fig. 3.6, is similar to that for the PET/PHBA system (Fig. 3.5), although the T_m for both the kinked homopolymer and the minimum melting composition are higher. This system, originally developed by ICI plc, has the advantage that all the monomers are relatively cheap. It is worthy of note that the liquid crystalline phase is stable up to compositions in which one third of the ring units can give kinks, and that mechanical properties such as axial stiffness are

Fig. 3.6 Phase diagram for the random copolymer system ⟨34⟩ based on hydroxybenzoic acid and equal amounts of isophthalic acid and hydroquinone. (Modified from [3.8], with acknowledgement to Dr W. A. MacDonald.)

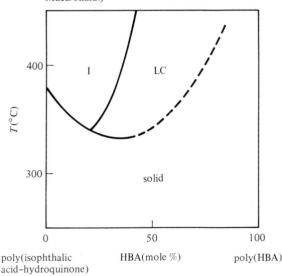

characteristic of rigid, extended molecules. Figure 3.7 illustrates how the conformation of the chain can adjust itself to minimize the influence of the kinked unit on the straightness of the molecule.

Units in which adjacent phenylene groups are joined through a single, tetrahedrally coordinated atom make very effective kinks. However, their incorporation by random copolymerization rapidly reduces T_m but also eliminates the mesophase. In particular bisphenol A, in concentrations as low as 15 mole %, appears able to destroy liquid crystalline behaviour of poly(hydroxybenzoic acid) when it is balanced by an equal molar concentration of terephthalic acid and randomly copolymerized ⟨35⟩. Bisphenol A is an attractive candidate as a modifier because it is cheap, although this advantage is of limited significance as it is only needed in small proportions. However, in

Fig. 3.7 Molecular model showing a part of a random copolyester containing equal proportions of isophthalic acid, hydroquinone and hydroxybenzoic acid units, ⟨34⟩, in a simulation of a possible conformation within a liquid crystalline phase. Rotations about the ester bonds have minimized the disruptive influence of the kinked isophthalic units (arrowed) on the straightness of the chain.

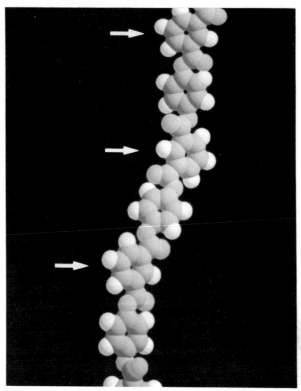

addition to readily suppressing liquid crystallinity, it also seems to have a particularly deleterious effect on the mechanical properties of the material.

Throughout this section, random copolymerization has been shown to have the effect of reducing the melting point of the polymers considered. However, it is appropriate to ask how a random molecule, without any long range periodicity, can contribute to a three-dimensional crystal lattice, in fact how it can crystallize at all.

Flory [3.17] has shown that a certain level of crystallinity is possible in random copolymer systems by virtue of the segregation of homopolymer sequences which occur by chance within the molecules. However, the crystallinities obtainable are low and even though schemes involving aperiodic crystallites can account for higher levels of crystallinity in some systems [3.18], it remains that crystallinities greater than 15% are unlikely in random copolymer systems with equal proportions of components. The development of solid state order in random copolymers, and its influence on mechanical properties, is discussed in further detail in later chapters.

A low level of crystallinity is not especially desirable in a polymer designed for mechanical performance at elevated temperatures: for, although the material will have to be processed above the melting point, the absence of significant crystallinity will mean that the mechanical properties are likely to be relatively dismal above the glass transition temperature, T_g. In fact, the useful service temperature limit is effectively determined by the glass transition. It is thus particularly unfortunate that many liquid crystalline copolymers seem to have lower T_g's than conventional isotropic equivalents of similar chemical make-up. Compare, for example, the first two polymers of Table 3.5 which contains T_g data from the work of MacDonald [3.19]. Polymer $\langle 36 \rangle$, which is neither crystalline nor liquid crystalline, has a T_g of 200 °C whereas polymer $\langle 37 \rangle$ which is liquid crystalline in the melt and probably between 5 and 15% crystalline below 290 °C has a T_g of 135 °C. The T_g's were measured by dynamic mechanical analysis.

The approach to increasing T_g in a liquid crystalline material is to design the molecule so that:

(a)   it contains the maximum number of kinked units commensurate with the retention of the liquid crystallinity over a usable temperature range

and

(b)   it contains linking units known to maximize T_g in conventional polymers, these units being comparatively resistant to rotation themselves.

Table 3.5 *The effect of modifying units on T_g* [3.19]

Molecule			T_g
<36>			200 °C
<37>			135 °C
<38>			140 °C
<39>			135 °C
<40>			180 °C
<41>			230 °C

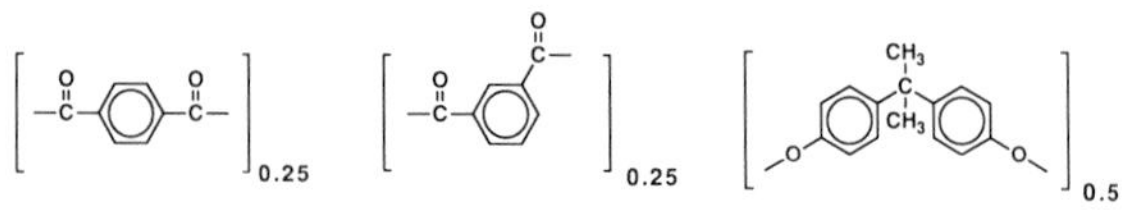

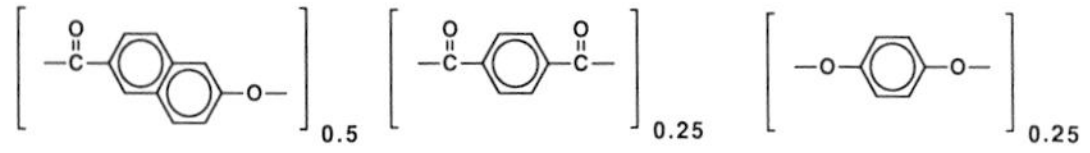

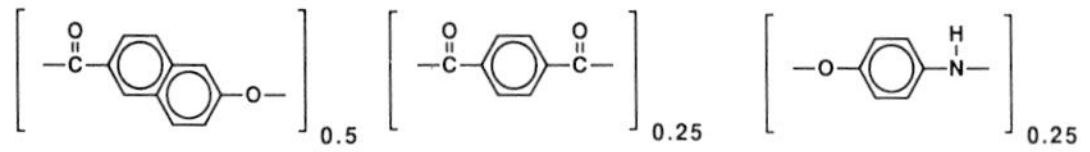

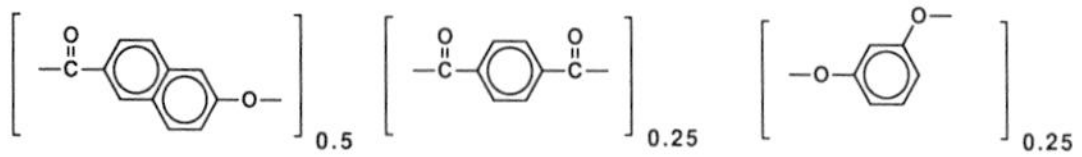

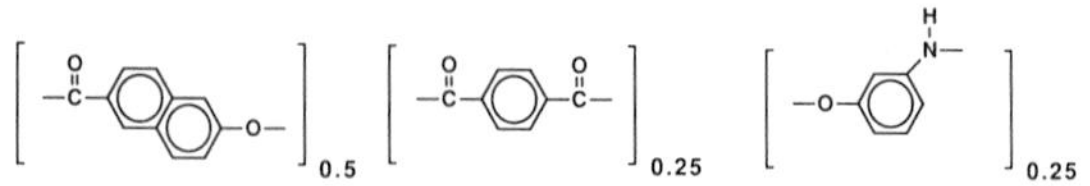

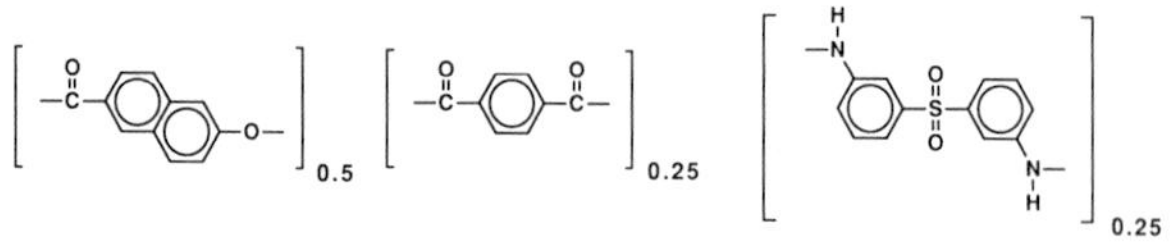

The importance of a *combination* of kinking with link stiffness is illustrated in the series of random copolymer molecules ⟨38⟩ to ⟨41⟩ (Table 3.5). *Para* (linear) amide links ⟨38⟩ or *meso* (kinked) units ⟨39⟩ in an ester environment have a comparatively modest effect in enhancing T_g, whereas a kinked *meta*-aminophenol unit ⟨40⟩, or even better, a combination of sulphone and *meta* amide groups in the 3,3′ diaminodiphenyl sulphone ⟨41⟩ serve to increase the T_g's to 180 °C and 230 °C respectively. Another aspect of introducing kinked units, is that the crystallinity is reduced to very low levels, and probably eliminated altogether. This loss is of little consequence with respect to the mechanical properties below T_g, but is an advantage in that the temperature range of mesophase stability is increased, and the temperature necessary for melt processing correspondingly reduced.

Table 3.6 *The influence of connection through flexible spacers on the transition temperatures of a small rod-like molecule*

	T_m	$T_{lc \to i}$
<42>	199 °C	223 °C
<43>	106 °C	121 °C

3.3.6 Modification by incorporating flexible chain segments in the backbone

The insertion of sections of flexible molecule to separate the mesogenic groups along a liquid crystal polymer backbone is another means of reducing the transition temperatures and exposing the thermotropic mesophase in a convenient temperature range. With this approach, the chemical periodicity of the molecule is preserved although the repeat distance is increased. The general configuration of such a polymer chain is shown schematically in Fig. 3.8. In Table 3.6 the transition temperatures of a specific example of such a polymer ⟨42⟩, in this case with a decamethylene flexible spacer, $-(CH_2)_{10}-$, are

Fig. 3.8 Schematic diagram of mesogenic groups connected together by flexible spacers.

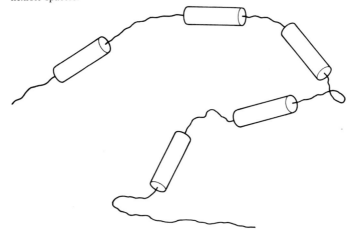

compared with those for a small molecule precursor ⟨43⟩ which has a similar mesogenic core terminated with ethoxy groups. This illustration emphasizes that it is possible to regard mesogenic polymers containing flexible spacers in the backbone in one of two lights. They can be thought of as small molecule mesogens connected together through identical flexible spacers, in this case the connection, or polymerization, leading to an increase in T_m of 93 °C and in $T_{lc \to i}$ of 102 °C, or otherwise the flexible spacers can be seen as reducing what would be inconveniently high transition temperatures of a polymer built from the mesogenic cores alone. Each approach has its merits, depending on the context, but the central issue remains the relationship between the thermotropic stability, as indicated by transition temperatures, and the nature and relative proportions of the mesogenic and flexible sequences.

Not surprisingly, both transition temperatures show a marked decrease with increasing length of flexible spacer. Figure 3.9 is a plot of both T_m and $T_{lc \to i}$ against the number of —(CH$_2$)— units in the spacer for molecules of the family illustrated by ⟨42⟩, which is the example

Fig. 3.9 Plot of T_m (●) and $T_{lc \to i}$ (○) against the number of —(CH$_2$)— units, m, in the flexible spacer for molecules of the type represented by ⟨42⟩. (Redrawn from [3.20].)

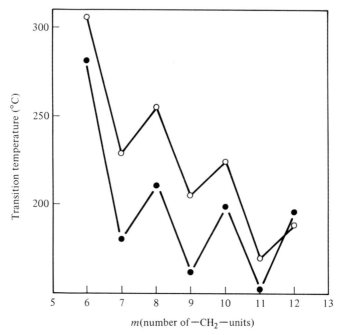

with $m = 10$. The zig-zag nature of both plots is a nice example of an effect known as the *odd–even effect*. Both melting and liquid crystalline → isotropic temperatures tend to be higher when there is an even number of —(CH$_2$)— units in the flexible spacer. This effect can be best understood by assuming the conformation of the methylene spacer to be all-*trans* (planar zig-zag). This conformation is straight and has the lowest energy of any conformation, and there is considerable evidence that it is the most likely conformation in both crystalline and liquid crystalline phases. However, as Fig. 3.10 illustrates, an all-*trans* sequence will only connect two mesogenic groups without introducing a kink if it has an even number of units. Note that the temperature range for mesophase stability decreases with increasing length of spacer. At $n = 12$ in this example, $T_{lc \to i}$ first dips below T_m and the liquid crystalline phase is lost. In the limit of $n \to \infty$, the molecule is polyethylene which is manifestly not mesogenic. However, it is noteworthy that polyethylene is predicted to have a liquid crystalline to isotropic transition at $-60\,°C$, which would be observable if crystallization did not intervene below $+130\,°C$.

Spacer segments other than polymethylene have been incorporated into mesogenic molecules. Polyethylene oxide gives much the same trends, although the mesophase is destroyed for spacer lengths in excess of eight chemical repeats. Polysiloxane spacers appear to be much more effective, per unit length, in reducing transition temperatures on account of their greater flexibility and larger diameter.

Fig. 3.10 For an extended polymethylene segment, an even number of —(CH$_2$)— units is necessary to keep the axes of the mesogenic units parallel.

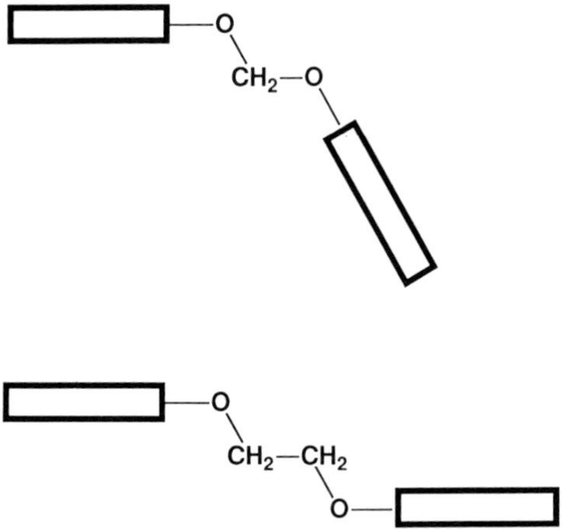

3.4   Lyotropic polymers
3.4.1   *Solvents and liquid crystal stability*

So far the molecules discussed have been thermotropic in that they show liquid crystallinity in a particular temperature range. *Lyotropic* systems differ in that the stability of the liquid crystalline phase is influenced by the addition of a low molecular weight solvent. Phase equilibria must therefore be considered as a function of solvent content as well as temperature, and phase diagrams are used to illustrate the equilibrium behaviour of the system.

An important consequence of solvent addition is that liquid crystallinity can be obtained with rigid-rod polymers which would not otherwise show the mesophase at temperatures below that of thermal decomposition. In other words lyotropy is a route to melting point reduction.

Solvent addition also reduces $T_{lc \to i}$. In fact, as rigid-rod polymers have an effective axial ratio well in excess of 6.4, an isotropic *melt* is not possible. It is therefore a method of achieving an isotropic phase. There is a critical concentration v_p' of polymer in the solvent, below which the solution is isotropic and above which a liquid crystalline phase forms.

Fig. 3.11 Plot showing the concentrations (v_p') of poly(p-phenylene-terephthalamide) (PPTA) ⟨44⟩ at which a liquid crystalline phase is first observed to form, as a function of the axial ratio of the molecule calculated from the molecular weight assuming perfectly straight chains. The data have been collected by Flory [3.21] and correspond to measurements in dimethyl acetamide–LiCl (○) and 85% sulphuric acid (●).

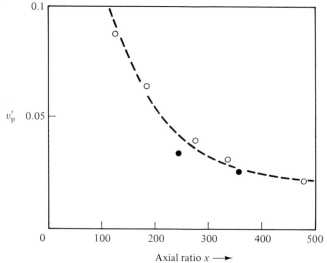

Axial ratio $x \longrightarrow$

This critical amount usually increases with the flexibility of the polymer chain (i.e. decrease in persistence ratio), but for systems in which lyotropy is the only route to liquid crystallinity, it is quite small, of the order of 5–10 %. The theoretical prediction of these stability criteria is an important aspect of the subject and is treated in detail in Chapter 4.

Figure 3.11 is a plot of the critical polymer concentration for liquid crystallinity, v'_p, against the nominal axial ratio of molecules of poly-(p-phenyleneterephthalamide) (PPTA) $\langle 44 \rangle$ covering a range of molecular

$\langle 44 \rangle$
poly(p-phenyleneterephthalamide)
PPTA

weights in two different solvents. The nominal axial ratio is calculated from the molecular weight assuming a completely straight and rigid backbone conformation. Note that an increasing axial ratio, obtained by using a polymer of higher molecular weight, leads to a decrease of the polymer concentration necessary to stabilize the lyotropic liquid crystalline phase, so that for axial ratios in excess of 400 the critical concentration is down to less than 3 %. In the case of molecules such as hydroxypropyl cellulose (HPC) $\langle 21 \rangle$, which are less stiff than PPTA, the effective axial ratio will be limited by a comparatively modest persistence length and thus be less than any value based on the full length of the molecule. HPC has a persistence ratio of 19 (Table 3.3), which would account for observations that its critical concentration for liquid crystallinity is always of the order of 0.3 or above, no matter how high the molecular weight.

3.4.2 *Phase diagrams*

A lyotropic system will show two liquid phases, an isotropic one at low concentrations of polymer and a liquid crystalline one at higher concentrations. However, as with any two-component system allowed to reach equilibrium, there will be a range of compositions in the transition region over which the two phases coexist. Within this region the composition of each phase remains constant (at constant temperature and pressure) but the relative amounts of each phase change from 100 % isotropic phase at the low (polymer) concentration limit, to 100 % liquid crystalline phase at the upper limit.

The combined effect of composition and temperature on phase stability is illustrated by plotting a *phase diagram*. Figure 3.12 shows the phase diagram for the lyotropic polymer–solvent system, PBLG–dimethylformamide (DMF). Above room temperature a comparatively narrow two-phase region is apparent, separating the areas of isotropic and liquid crystalline stability. It is sometimes referred to as the *biphasic chimney*. Within this region the compositions of the coexisting phases, one isotropic and one liquid crystalline, are very similar. At lower temperatures there is a pronounced increase in the composition range over which two phases coexist to the extent that the isotropic and ordered phases are distinct over much of the width of the diagram. The reason for this marked change in behaviour is associated with the chemically driven segregation of the two components which manifests itself at lower temperatures where the term involving the entropy of mixing which encourages miscibility ($T\Delta S_{mix}$) has less effect.

The general form of this phase diagram was predicted by Flory in 1956 [3.23] using a two-component lattice model, nearly 20 years before the experimental observations of Miller *et al.* on PBLG [3.22]. Figure 3.13 shows the theoretical diagram for rods of axial ratio 100. Note the similarity in form to the experimental data of Fig. 3.12, but that the vertical scale is in terms of the *Flory–Huggins interaction parameter χ* (plotted here as minus χ upwards) rather than temperature. A reduction in χ will have the effect of discouraging the aggregation of rods, which is an equivalent effect to an increase in temperature. The

Fig. 3.12 Phase diagram of the system PBLG $\langle 22 \rangle$ (molecular weight 310 000) and dimethyl formamide (DMF). Note the biphasic 'chimney'. [3.22]

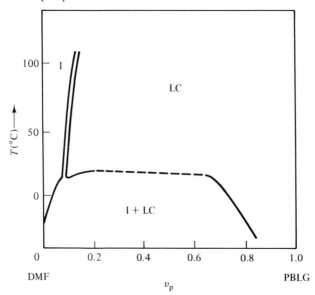

theory of phase stability as well as relationship between χ and T are examined more closely in Chapter 4. The theoretical phase diagram does not consider the possibility of the existence of other phases in the more concentrated regime, such as the crystalline polymer phase, and indeed much of the experimental work reported (as typified by Fig. 3.12) has focused on the more dilute part of the phase diagram. However, other phases have to be considered and the full phase diagram is likely to be rather more complex than that first proposed by Flory.

A practical way of adjusting χ is to use a ternary system with two diluents which are completely miscible in each other and yet have different thermodynamic efficiencies as solvents (i.e. give different values of χ). Figure 3.14 shows one part of a ternary diagram for PBLG in mixtures of DMF and methanol [3.24]. The addition of methanol decreases χ and when present in volume fractions of 0.12 or more, it stabilizes the wide biphasic region with respect to the narrow chimney.

While noting the general accord of the experimental PBLG–DMF

Fig. 3.13 Phase diagram of the type first calculated by Flory for rods of axial ratio $(L/d) = 100$. The fact that the region under the shallow dome consists of two liquid crystalline phases of different composition is made clearer by designating compositions just to the left of the dome LC′ and to the right, LC″. χ is the Flory–Huggins interaction parameter (see Chapters 2 and 4).

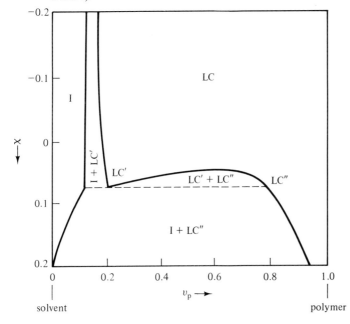

diagram (Fig. 3.12) with theoretical predictions, it is rewarding to look at it a little more closely. When cooling across the $I \to I + LC$ boundary, the onset of the new phase can be detected by the appearance of a birefringent microstructure, while the loss of fluidity as the regions of liquid crystalline phase become continuous, may not commence until some 10 °C to 15 °C lower. However, when cooling more concentrated ordered solutions into the broad biphasic region, the exact transition temperature is difficult to determine. There is a significant rise in viscosity, and in PBLG–DMF at least there is also a marked reduction in the second virial coefficient, indicating a decrease in the solvent efficiency of the DMF. The dashed phase boundary in Fig. 3.12 reflects this difficulty, and it is not until recently that the predicted 'domed' region of the diagram in which two liquid crystalline phases coexist has been confirmed experimentally for PBLG in benzyl alcohol [3.25, 3.26].

The curvature of the chimney towards higher concentrations with increasing temperature apparent in Fig. 3.12 represents the increase in molecular flexibility with temperature. Papkov [3.27] has suggested that if higher temperatures were to be accessible – they are not for reasons of diluent volatility and polymer instability – then the biphasic region would eventually connect with the pure polymer axis at a temperature which could be identified with the upper thermotropic transition ($T_{lc \to i}$),

Fig. 3.14 Portion of an experimental ternary phase diagram for PBLG with a nominal axial ratio of about 350. Dimethyl formamide is a solvent and methanol a non-solvent. Note the biphasic chimney and the same general form as seen in the binary systems as a function of temperature. (Redrawn from data in [3.24].)

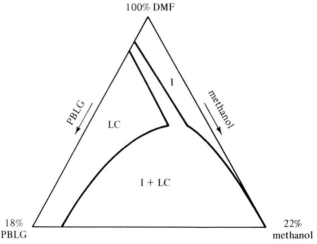

100% DMF

PBLG

LC

I

methanol

I + LC

18%
PBLG

22%
methanol

where the molecule is sufficiently flexible to give an isotropic phase in the pure polymer. His schematic diagram is shown in Fig. 3.15.

3.4.3 *Phase equilibria with polymer and crystal solvates*

As indicated above, neither the theoretical phase diagram of Flory (Fig. 3.13) nor the experimental diagram for PBLG–DMF of Miller (Fig. 3.12), shows the possibility of equilibria between the polymer rich phases and the crystalline polymer itself, nor do they provide a ready explanation for the reported formation of a gel on cooling into the broad biphasic region and the electron microscopy observations of a fibrous network structure. Figure 3.16 shows the gel melting temperatures superimposed on phase diagrams for the system PBLG–benzyl alcohol (BA) in which two distinct gel forms have been identified depending on the route of preparation. It has been suggested that there is an extrapolation to a transition at 135 °C for the pure

Fig. 3.15 Schematic phase diagram of a lyotropic system showing the possible curvature of the biphasic chimney towards the polymer axis at high temperatures. [3.27]

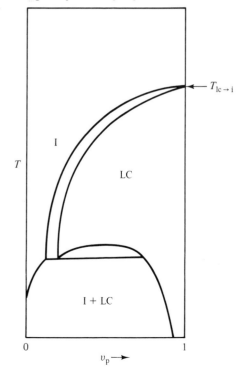

polymer and this has been identified with the collapse of the stacking structure of benzene rings in the polymer [3.28]. The high melting point of the gel compared with the biphasic liquid crystalline transition indicates that it contains new phases not shown in the basic phase diagram. Recently [3.25], it has been suggested that the actual structure of the gel derives from phase separation (into isotropic–liquid crystal or liquid crystal–liquid crystal phases depending on the value of χ) which is then stabilized by the existence of crystals – considerable crystalline-type order can be detected. Sasaki *et al.* [3.29] identify the higher of the two melting lines shown in Fig. 3.16 with the demise of a gel which was mechanically stabilized by the crystalline phase, while the gel which melts at the lower line contains entities of crystal solvate rather than crystallites. (A crystal solvate is a polymer crystal with a fixed proportion of solvent incorporated within the regular lattice. Solvates are also observed in PBA and PPTA systems with sulphuric acid, and their structures have been determined in some detail.)

Fig. 3.16 Plot of apparent 'gel melting' temperatures as a function of polymer concentration in the solvent for the system benzyl alcohol (BA)–PBLG. [3.29]

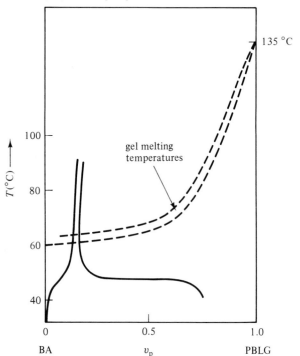

The question remains as to the form of the phase diagram which properly represents the equilibria between the liquid crystalline phases and the crystal and, if present, the crystal solvate. A possible diagram is reproduced in Fig. 3.17. The solvate phase is deemed to contain 25 % solvent, while there are temperature invariant transitions at T_1 and T_2. At T_1, the liquid crystalline phase of composition v_p''' interacts with the stoichiometric amount of crystalline polymer phase to form a crystal solvate, while at T_2 the liquid crystalline phase of composition v_p'' separates into the crystal solvate phase and an isotropic phase of composition v_p'.

3.4.4 *Practical lyotropic polymer systems*

The phase diagram (Fig. 3.17) illustrates nicely how the addition of solvent molecules to the pure polymer can be viewed as a means of reducing the melting point of the polymer, and thus classed with the various methods of reducing T_m already described for thermotropic materials. In fact there are several systems in which the methods of random copolymerization are not completely successful in reducing the melting temperature sufficiently to give liquid crystalline stability at acceptable processing temperatures. Notably, these include the aromatic polyamides where the hydrogen bonding between chains

Fig. 3.17 Schematic plot of a lyotropic phase diagram which shows a crystal solvate phase (CS) and the pure polymer crystal (C). (After [3.27].)

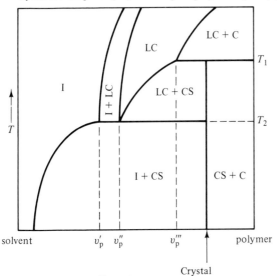

tends to elevate the melting point, and rigid-chain aromatic heterocyclic molecules. Several of the cellulosic derivatives also fall within this class, though mainly on account of their low degradation temperatures.

The aromatic polyamides (polyaramids): poly(*p*-benzamide) (PBA) ⟨23⟩ and poly(*p*-phenyleneterephthalamide (PPTA) ⟨44⟩ show lyotropic but not thermotropic behaviour. They have been widely studied because of their potential for the manufacture of very strong low density fibres. The best known commercial products based on PPTA are Kevlar (Du Pont) and Twaron (Akzo). These fibres are spun from lyotropic dopes consisting, typically, of 20% polymer in 99.8% sulphuric acid. The phase is a solid at room temperature so it is extruded at between 70 and 90 °C into air and then immediately passed through a coagulant bath of water at 5 °C. At this stage the acid diffuses out of the fibre which crystallizes in a highly oriented condition. The process is referred to as 'dry jet wet spinning'. The fibres are strong, stiff, chemically resistant and retain much of their properties up to 300 °C (see Chapters 7 and 8 for further details of the preparation and properties of these materials). Various grades are manufactured which fine tune the properties to applications such as fibre composites, ropes and extra tough fabrics.

The aromatic heterocyclic polymers are typified by the polybenzazoles such as poly(*p*-phenylenebenzobisthiazole) (PBTZ) ⟨45⟩ and poly(*p*-phenylenebenzobisoxazole) (PBO) ⟨46⟩. The most useful

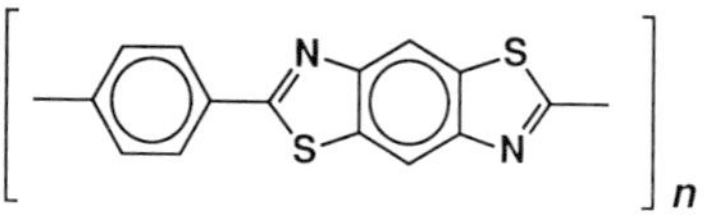

⟨45⟩   PBTZ

poly(*p* - phenylene benzobisthiazole)

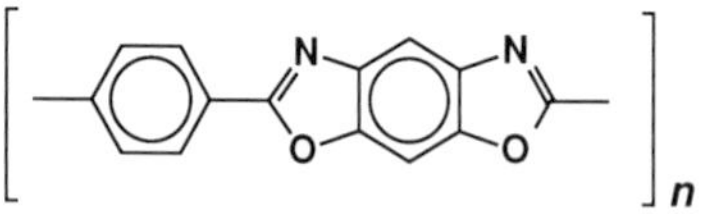

⟨46⟩   PBO

poly(*p* - phenylene benzobisoxazole)

molecules are *trans* PBTZ and *cis* PBO, and it is these stereoisomers which are represented in the molecular diagrams. Polymerization is carried out in polyphosphoric acid and the fibres or films are dry jet wet spun either from a 5% lyotropic solution in this solvent or from a new 10% solution in methane sulphonic acid with a few per cent chlorosulphonic acid. The solvent is removed in a coagulation bath to give a highly aligned product with exceptional mechanical properties [3.30]. The molecules are very rigid, the only rotational freedom being around bonds which are closely parallel to the chain axis, and for this reason they are sometimes referred to as 'linkageless'. Their development has been driven by military requirements for a non-conducting fibre with properties which approach those of carbon, and more recently the same polymers have become the centre of attention in the development of non-linear optical materials (Chapter 8).

The potentially important development of *molecular composites*, has been made with these polybenzazoles very much in mind as they approximate to rigid rods and thus have a very low v'_p. The concept of a molecular composite envisages a three-component system consisting of a rigid-rod polymer, a flexible-coil (conventional) polymer, and a solvent common to both. Ideally, the rigid-rod component would stabilize a lyotropic mesophase in which the role of the solvent is played by both the conventional polymer and the low molecular weight solvent. The difficulty to be overcome concerns the phase stability of the system. For while dilute isotropic solutions of both polymers in the common solvent are single phase, once the concentration is increased into the mesophase region, phase separation occurs with partition of the vast majority of the rigid-rod molecules into the liquid crystalline phase and the flexible molecules into the isotropic one. Such partition defeats the primary objective of a molecular composite, which is to achieve a *molecular* dispersion of the rigid rods in an entangled melt of flexible chains. However, if the solvent is removed sufficiently rapidly, the molecular segregation has insufficient time to develop over significant distances. So, although complete molecular dispersion is probably never achievable, the groupings of rigid-rod molecules which do occur are sufficiently small to give a range of materials with potentially important properties [3.31]. Systems based on PBTZ, methane sulphonic acid and either Nylon or a non-linear aromatic molecule [3.32] as the flexible component are currently under development.

3.5 Polymers with mesogenic side chains

3.5.1 *Molecular architecture*

The idea behind side-chain liquid crystalline polymers is the grafting of mesogenic side groups onto a suitable flexible polymer molecule. Figure 3.18 shows the schematic principle. In fact, diagram (*b*) is an ideal state which would only be achieved if the backbone conformation geometry was perfectly matched to the packing requirements of the side chains. In practice, the connection of mesogenic groups by direct attachment to a common backbone appears to prevent the formation of a mesophase; it also greatly increases the glass transition temperature of the polymer and prevents its crystallization.

Fig. 3.18 Mesogenic side groups need to be joined to a flexible backbone through decoupling units if they are to exhibit liquid crystalline order (see text).

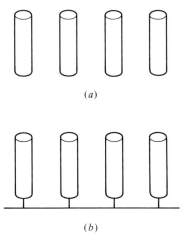

(*a*)

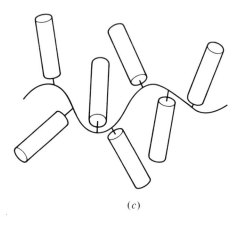

(*b*)

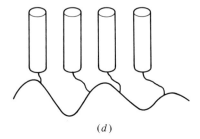

(*c*)

(*d*)

Where some liquid crystalline structure is apparent in the 'glass', it is invariably permanently lost on heating above T_g, being merely a non-equilibrium memory of the original small-molecule mesophase. These effects occur because the tendency of the backbone to adopt its preferred disordered conformation is antagonistic to the mesogenicity of the pendant side groups. A more realistic picture of a polymer molecule with directly attached mesogenic side chains is given by Fig. 3.18(*c*). The key to making side-chain liquid crystalline polymers is the incorporation of a flexible connecting link between the mesogenic group and its attachment point to the backbone. The rigid groups are thus sufficiently decoupled from the perturbing influence of the backbone to pack as a mesophase. Figure 3.18(*d*) illustrates the point. The flexible links are typically polymethylene, and it has been found that links of two or four $-(CH_2)-$ groups are often sufficient to achieve the required decoupling. There is a vast number of possible combinations of backbone, flexible link length and mesogenic group, although backbones are usually polyacrylates, methacrylates or siloxanes, e.g. molecules $\langle 47 \rangle$ and $\langle 48 \rangle$.

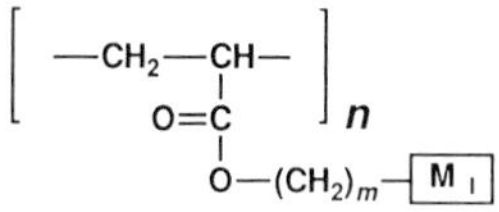

$\langle 47 \rangle$

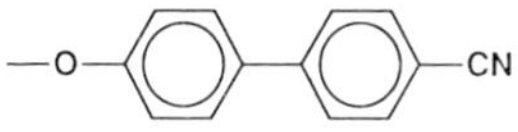

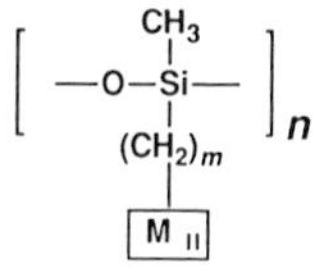

$\langle 48 \rangle$

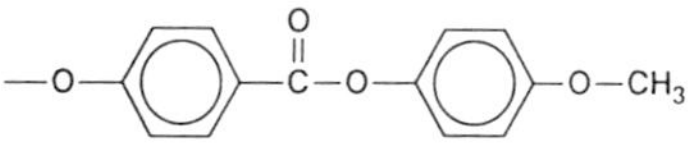

3.5.2 *Transition temperatures*

The solid curves of Fig. 3.19 show T_g and $T_{lc \to i}$ as a function of the number of CH_2 units in the flexible links for an acrylic polymer such as $\langle 47 \rangle$ with mesogenic side chains M_I. The dashed curves represent T_m and $T_{lc \to i}$ for the side chains on their own, i.e. small molecules of structure M_I. Note that the mesogenicity of the small molecule is stabilized by the flexible tail, and that attachment to the backbone greatly increases the stability range of the mesogenic phase. Crystallization is prevented so that the lower bound of stability becomes the glass transition temperature, whereas $T_{\text{lc}\rightarrow\text{i}}$ is substantially increased as a consequence of the reduced entropy of the mesogenic units in the isotropic phase due to their attachment to the backbone. The system chosen for Fig. 3.19 shows an increase in $T_{\text{lc}\rightarrow\text{i}}$ with increasing m, for the mesogenic sequences both as isolated molecules and as side groups. This behaviour is not universal to all systems and examples could be given, particularly where the mesogenic part of the molecule is longer, where the trend is in the opposite direction.

The influence of the molecular weight of the flexible backbone on the transition temperatures is only significant for comparatively short molecules. Data for T_{g} and $T_{\text{lc}\rightarrow\text{i}}$ from [3.34] for polymer $\langle 48 \rangle$ with side chains M_{II} and $m = 3$ are drawn in Fig. 3.20 as a function of the degree of polymerization, n. For chains longer than eight repeat units there is little effect and it is interesting that the mesophase itself is not stabilized until the chain is three units long. Many side-chain liquid crystalline polymers form smectic structures which are discussed in detail in Chapter 5. However, it should be noted that the nematic phase

Fig. 3.19 Plot showing the effect on the transition temperatures, $T_{\text{lc}\rightarrow\text{i}}$ and T_{g} (solid curves), of different numbers of carbon atoms, m, in the flexible spacer joining the backbone $\langle 47 \rangle$ to the mesogenic side group M_{I}. The dashed curves are $T_{\text{lc}\rightarrow\text{i}}$ and T_{m} for the side groups (mesogenic part and flexible spacer as tail) on their own. (Data collected by Shibaev [3.33].)

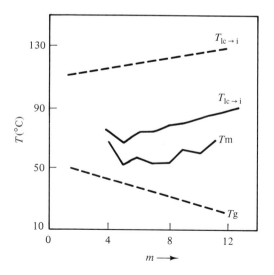

is generally favoured for structures where the mesogenic group and flexible link are short, the backbone is comparatively flexible, and there is only a short substituent group (if any) at the outer end of the side chain. It seems a rule of thumb that if the rigid group is sufficiently anisotropic to form a small-molecule nematic phase, then it will give a smectic phase upon polymerization.

3.5.3 *Combined main-chain and side-chain polymers*

An additional richness of structures and phases can be found for molecules where both the backbone and the pendant groups are potentially capable of alignment, as may occur for instance in the case of combined main-chain and side-chain polymers included in Fig. 3.1. For such molecules, as has been discussed by Warner [3.35, 3.36], two order parameters are required to describe the state of the polymer – one for the backbone and one for the side chains. These two parameters may or may not have the same sign – in this context a plus sign implying alignment along an external field. Thus there are three possible nematic phases, shown in Fig. 3.21, which can be identified as $(+,-)$, $(-,+)$, and $(+,+)$. The backbone and side chains may

Fig. 3.20 Plot showing the influence of the molecular weight of the backbone, expressed as the degree of polymerization, on transition temperatures of polymer $\langle 48 \rangle$ with side chains M_{11} and $m = 3$. (Redrawn from [3.34].)

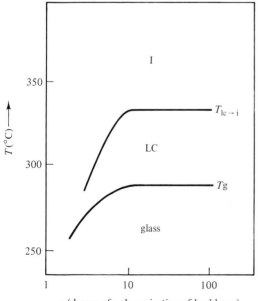

n (degree of polymerization of backbone)

themselves possess a coupled interaction which can be temperature dependent, and thus it is predicted that there will be transitions between the different phases as a function of temperature. Phase diagrams have been drawn for different types of coupling [3.36], but as yet there is no experimental evidence to support the form of these.

3.5.4   Copolymers

If two different types of side chain are connected to the backbone in random sequence, the properties of the liquid crystalline polymer will depend on their relative proportions. Where one of the groups is small and non-mesogenic, the mesophase can be preserved down to surprisingly small proportions of the mesogenic group (perhaps 20 mole %). On the other hand larger non-mesogenic side groups effectively screen the mesogenic ones from each other, and the mesophase is destabilized at much lower levels of copolymerization. In the case where two different side groups are both nematogens then T_g, and to some extent $T_{lc \to i}$, vary smoothly between the composition limits, although there may be a composition at which there is a transition between mesophase types, e.g. nematic to smectic. Random co-polymerization in side-chain materials opens up the possibility of multifunctionality, where particular groups are designed for a particular purpose. That may simply be the stabilization of the liquid crystalline phase, or the introduction of particular electro-optical effects which could involve either non-linear responses or coupling into other fields such as mechanical or magnetic fields. The molecular engineering of side-chain polymers for device application is described in Chapter 8.

Fig. 3.21 Three possible states of alignment for a molecule consisting of a mesogenic backbone and mesogenic side groups connected through a flexible spacer. In (*a*) the side chains are aligned along the (vertical) field and the main chains normal to it (to give positive and negative order parameters respectively), in (*b*) the alignments are reversed, while in (*c*) both backbone and side groups are aligned parallel with the vertical field. [3.36]

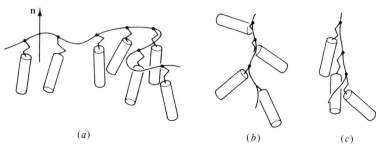

(*a*)   (*b*)   (*c*)

4 Theories of liquid crystalline polymer systems

4.1 Introduction

In the previous chapter, the design of mesogenic molecules was examined in some detail with a view to understanding the relationship between chemical structure and liquid crystalline stability. However, just as it is important to relate properties to the chemical detail, it is of equal significance to be able to determine the stability of a mesophase on the basis of simple schematic models of the molecules – the simplest model of these is likely to be the most attractive to the theorist and it will have the great advantage of exposing the essence of any situation. For, in finding just those few features which are really necessary to make the model work, one is at the same time identifying them as *the* key elements.

The theories described in this chapter focus on one issue; that is whether a given type of molecule will lead to an isotropic or a liquid crystalline state. The transition between the two states is examined in detail as a function of temperature and/or solvent content. Theory to predict either the crystalline melting point or the glass transition temperature is not included, for the main reason that it has yet to be developed much beyond the empirical stage. The molecular theory of liquid crystals will be discussed first. Our understanding of the parameters governing the onset of liquid crystallinity and the form of phase diagrams for lyotropic systems owes much to the pioneering work of Onsager [4.1] and Flory [4.2]. Some later developments allow departures from ideality to be taken into account, specifically a decrease in the rigidity of the rod and molecular weight polydispersity.

Of the models, Flory's lattice theory is developed in most detail, for not only can it be described at the modest mathematical level of this text, but it has been extensively applied to many aspects of liquid crystalline polymers. In several respects the lattice model builds on the earlier ideas of Onsager, and the reader is referred to de Gennes [4.3] for a simple comparison of the two contributions. There is also a very useful review written by Flory in 1984 [4.4] which covers most relevant topics in the context of the lattice theory and its later enhancements.

The mean field theory, developed by Maier and Saupe [4.5, 4.6], considers anisotropic intermolecular attractions due to long range dispersion forces. Although it has been very successful in accounting for the structure of small-molecule liquid crystals, it is not so readily applicable to polymers. Nevertheless, some recent work has demonstrated the relevance of the mean field approach to polymers by providing a modifying orientation dependent term in extensions of the lattice treatment.

4.2 Steric theory of liquid crystals

In its simplest form, the steric theory models liquid crystalline molecules as rigid rods. The underlying rule is that they cannot interpenetrate and thus there is no possibility of two molecules occupying the same region of space. It is analogous to hard sphere models of simple crystal structures, in that *attractive* intermolecular forces are not taken into account in any specific way. In this respect steric models are in marked contrast to the Maier–Saupe theory; however they have proved themselves readily applicable to polymeric systems.

Onsager [4.1] and Ishihara [4.7] were the first to approach the excluded volume problem for pairs of asymmetric molecules. Their methods are comparatively complex. Also, for molecules of high asymmetry, where coefficients beyond the second in the virial expansion are required, the expressions become prohibitively difficult to solve numerically and the virial series does not converge for the concentrations required to form a stable nematic phase.

The central objective in any steric approach is to determine the number of ways of arranging a population of rods at a particular concentration in a given volume. It is necessary, not only to specify the axial ratio of the rods, but also their orientation distribution with respect to a defined principal axis, or *director*. The significance of calculating the number of arrangements, known as the partition function, Z, is that it is directly related to the entropy of the system, S, via the Boltzmann relation;

$$S = k \ln Z \tag{4.1}$$

while S, in turn, is a major factor in determining the thermodynamic stability of the system as expressed by the free energy. k is the Boltzmann constant.

In 1956 Flory [4.2] introduced his lattice model, and this has proved to be the most tractable of the steric theories. It can be used with little approximation in the regime of interest, where the average misorientation of rods away from the director is comparatively small. A given volume of material is divided into many identical cubic lattice cells, which are either completely occupied by a part of a rod molecule, or are empty. In fact, the vacant cells are considered to contain solvent molecules, so the model is relevant to lyotropic systems, with the pure thermotropic polymer but a limiting case.

Figure 4.1 gives the essential parameters of the model. The rod is assumed to comprise x identical units, each of which occupies one cell of the lattice (i.e. the side of each cell is equal to the rod diameter); a cubic lattice is chosen with one of its principal axes aligned with the director. Any rod molecule, inclined at an angle θ to the director, is broken down into a number of sequences, each of which lies parallel to the director. The parameter y is the number of such sequences and is a measure of the disorientation of the rod with respect to the director. The length of each sequence is thus given by the ratio x/y. Within the cubic lattice, the rod length expressed as x is identically equal to the *axial ratio*, since this is defined as rod length/diameter and the diameter is set equal to unity by virtue of the size of the cell chosen.

Flory chose to divide the partition function into two components: a steric or combinatory part, Z_{comb}, and an orientational part, Z_{orient}, so that:

$$Z = Z_{comb} Z_{orient} \qquad (4.2)$$

Fig. 4.1 (*a*) Rod-like molecule oriented at an angle θ to some preferred direction. (*b*) Subdivisions of the rod into square units on a square lattice. The total length of the rod is x units, and there are y straight sequences each of length x/y. The director is horizontal.

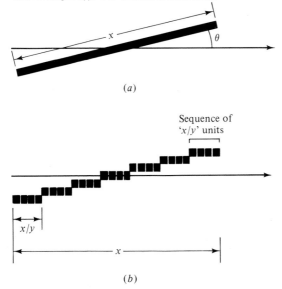

(*a*)

(*b*)

4.2.1 *The combinatory partition function, Z_{comb}*

This component of the partition function describes the number of ways of arranging the positions of n_p identical rod molecules on the lattice, when the orientation of each rod (as defined by its misorientation, y, with respect to the director, and its orientation *about* the director) is fixed. Consider adding the jth rod at some fixed orientation to the lattice already containing $(j-1)$ rods. If it can be added in v_j different positions, then Z_{comb} is given by the products of the v's for each of the rods successively added. Hence:

$$Z_{comb} = \frac{1}{n_p!}\prod_{j=1}^{n_p} v_j \qquad (4.3)$$

where n_p is the total number of identical rod-like molecules. The factor $1/n_p!$ is necessary because the molecules are indistinguishable, and all permutations of the molecules are equivalent. Even before calculating v_j explicitly it is clear that Z_{comb} must decrease as the disorder increases since as the rods misalign, there is an increased likelihood of steric clashes. Hence this part of the partition function is maximized for complete order, i.e. all molecules lying parallel to the director.

The next step is to determine v_j. This quantity will be the product of three terms: the number of lattice sites available for the first unit of the first straight sequence of the rod, A_j; the probability that the first sites required by each of the remaining sequences be vacant, P_j; and the probability that the second and subsequent sites of each sequence are also vacant, N_j. The first term is given by:

$$A_j = (N - x(j-1)) \qquad (4.4)$$

where N is the total number of lattice sites.

For the following units of the first sequence of the jth molecule, the fact that the site required by the first unit is already known to be vacant must be taken into account. Thus if the second unit site is occupied, it must be by the first unit of a sequence. There are $\bar{y}(j-1)$ such sequences, where

$$\bar{y} = \frac{1}{n_p}\sum y n_{py} \qquad (4.5)$$

is the mean value of y, and n_{py} is the total number of rods with misorientation y. If the site is vacant, it must be one of $N - x(j-1)$ vacancies. The probability that the second site required by the sequence is vacant is therefore a conditional probability, N_j, and is the ratio of the number of vacancies to the total number of vacant sites plus sites filled with the first unit of a sequence

$$N_j = \frac{N - x(j-1)}{N - x(j-1) + \bar{y}(j-1)} = \frac{N - x(j-1)}{N - (x - \bar{y})(j-1)} \tag{4.6}$$

This conditional probability is the same for all remaining units of this sequence, and for all units bar the first of the subsequent sequences. There are thus $x - y_j$ such units, and the probability that they will all be vacant is

$$N_j^{(x-y_j)}$$

For the first segment of the next sequence the prerequisite that the preceding site in the row be vacant, is not required. Thus for this site it is simply the probability of a vacancy, P_j, that is required. P_j is equal to the volume fraction of vacancies:

$$P_j = \frac{N - x(j-1)}{N} \tag{4.7}$$

There are $(y_j - 1)$ first units of sequences subsequent to the first, and therefore the probability that all will be vacant will be

$$P_j^{(y_j-1)}$$

The total number of accessible positions for the jth chain on the lattice, v_j, is therefore given by:

$$v_j = A_j N_j^{(x-y_j)} P_j^{(y_j-1)} \tag{4.8}$$

For a macroscopic system this can be reduced to:

$$\frac{[N - x(j-1)]! \, [N - (x - \bar{y})j]!}{(N - x_j)! \, [N - (x - \bar{y})(j-1)]! \, N^{(y_j-1)}} \tag{4.9}$$

and by substitution into (4.3) this equation yields for Z_{comb}

$$Z_{comb} = \frac{(n_s + \bar{y} n_p)!}{n_s! n_p! (n_s + x n_p)^{n_p(\bar{y}-1)}} \tag{4.10}$$

where $n_s = N - x n_p$ is the number of vacant sites, available to solvent molecules. In equation (4.10) both the numerator and the denominator increase with the misorientation, $\bar{y}$, but the increase is faster for the denominator. Hence Z_{comb} decreases as misalignment increases, as concluded above on the basis of purely intuitive reasoning.

4.2.2   *The orientational partition function, Z_{orient}*

The combinatory partition function just considered represents the number of positional arrangements of an assembly of rod-like molecules, each rod being of fixed orientation. The orientational partition function takes into account the many additional arrangements which are possible when a range of orientational options are assigned to each rod. For a rod at a given misorientation with respect to the director, defined by y, there will be a range of orientations in a conic distribution around this axis. For n_{py} rods at a given misorientation angle, there are

$$\left(\frac{\omega_y n_p}{n_{py}}\right)^{n_{py}} \text{ ways of arranging them}$$

In this equation, ω_y is the solid angle associated with a single value of y, and it is apparent that it will increase with increasing y. In any real system there will be a distribution of misorientation angles, and therefore a range of y values. The total number of orientational arrangements is therefore dependent on:

$$Z_{orient} = \prod_y \left(\frac{\omega_y n_p}{n_{py}}\right)^{n_{py}} \tag{4.11}$$

This equation gives the exact form of Z_{orient}. However, to proceed it is necessary both to know the relationship between ω_y and y and also the distribution of orientations as a function of y. These exact calculations have been performed by Flory and Ronca [4.8] using numerical methods. However, in order to keep the development as uncluttered as possible we will use the approximation of Flory's original 1956 paper which allows an analytical solution. The polar angles θ of the rods are assumed to be uniformly distributed over a range of angles $0 < \theta < \theta'$, with no rod possessing a polar angle greater than θ'. Taking random orientation as the reference state [4.4], Z_{orient} can be shown to reduce to:

$$Z_{orient} \sim \left(\frac{\bar{y}}{x}\right)^{2n_p} \tag{4.12}$$

4.2.3   *Behaviour of the partition function*

With expressions for Z_{comb} and Z_{orient} now defined, we can turn to examine the behaviour of their product, the total partition function. For small axial ratios and/or concentrations, Z is dominated by Z_{orient}. In this case, stability of the phase will increase with increasing misorientation, and the isotropic state will be the equilibrium state. Where Z_{orient} and Z_{comb} are more evenly balanced, the fact that the former increases with misorientation and the latter decreases leads to a maximum in Z and therefore phase stability will be optimal at some intermediate value of orientational disorder which is itself a function of concentration determined by n_p. In fact, a more convenient measure of concentration is the volume fraction of rods,

$$v_p = \frac{xn_p}{n_s + xn_p} \tag{4.13}$$

and results from now on will be expressed as a function of this variable.

Figure 4.2 shows the form of the partition function, Z, plotted as its *negative* logarithm, and normalized to the value of Z for perfectly oriented rods. The equation for $\ln Z$ is a form of the product of (4.10)

Fig. 4.2 Plot of partition function, Z as its negative logarithm, for rods of axial ratio 75, as a function of the degree of misalignment ($\bar{y}/x$) for different polymer concentrations, v_p. The lower the curve the greater the stability of the phase, and stable combinations are marked for the different compositions by ▲ and the metastable ones by △.

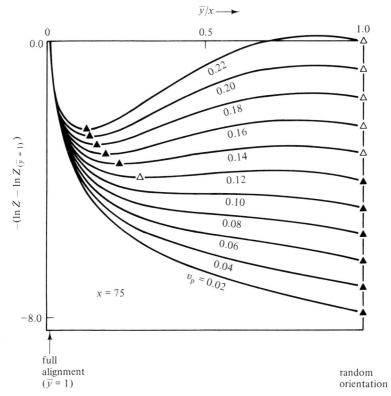

and (4.12) simplified by using Stirling's approximation for the logarithm of a factorial:

$$-\ln Z = n_s \ln v_s + n_p \ln (v_p/x) + n_p(\bar{y}-1)$$
$$- (n_s + \bar{y}n_p) \ln (1 - v_p(1 - \bar{y}/x)) - 2n_p \ln (\bar{y}/x) \quad (4.14)$$

The family of curves corresponds to a selection of values of the concentration, v_p, all with the same axial ratio of 75. The plot shows clearly that at low values of concentration, the minimum value of $-\ln Z$ and thus of free energy occurs at maximum disorder and the stable phase is thus isotropic ($\bar{y} = x$). As the concentration increases, a second minimum begins to appear at an intermediate degree of misorientation. It is first apparent at $v_p = 0.12$, and is the lowest minimum (solid triangle) at $v_p = 0.14$ and above. It is important to appreciate that the appearance of a minimum in the plot is not of itself sufficient to lead to the formation of a stable partially ordered state, for unless the minimum is lower than the value for the disordered state, the phase will be metastable as indicated by the unfilled triangles on Fig. 4.2.

It is useful to identify a critical value of v_p designated v_p^*, at which the minimum corresponding to a partially ordered phase first appears. For $x = 75$ $v_p^* = 0.11$. Flory [4.2] derived a useful semi-empirical formula relating the critical volume fraction v_p^* to the axial ratio x. It is:

$$v_p^* = (8/x)(1-2/x) \quad (4.15)$$

Although this expression is frequently used to estimate the critical volume fraction at which a stable ordered phase first forms, this value will be an underestimate since it corresponds to the appearance of a minimum in the $-\ln Z$ plot, rather than the lowest minimum. In practice, however, the value is likely to be reasonably accurate because the approximations underlying the 1956 theory themselves lead to an overestimate of v_p.

It is interesting to note that when $v_p^* = 1$, equation (4.15) gives a limiting axial ratio of 4, while the exact theory of Flory and Ronca [4.8] gives the widely quoted value of 6.4. These numbers can be interpreted as being estimates of the *minimum* axial ratios necessary before rigid-rod molecules can form a *thermotropic* liquid crystalline phase.

4.2.4   *Prediction of phase diagrams*

So far the theory has shown that there is a critical polymer volume fraction v_p^* (or axial ratio) above which a metastable ordered state exists, and that v_p^* decreases as the axial ratio increases (equation 4.15). The next step is to calculate the partition function for a range of concentrations and axial ratios, to generate a phase diagram.

At constant temperature $\ln Z$ is proportional to free energy as long as the solution is athermal. This means that the enthalpy of mixing and thus the Flory–Huggins interaction parameter χ (Chapter 2), to which it is related, are zero. On this basis, the free energies relative to the fully ordered state of both the isotropic and anisotropic phases, given by the minima on Fig. 4.2, are plotted against composition in Fig. 4.3, for the axial ratio of 75. The usefulness of free energy/composition plots is that they form a good framework for determining the range of biphasic equilibrium. It is a consequence of the phase rule that, in any phase diagram where composition is a variable, then regions of single phase stability are always separated by a region in which the two phases coexist in equilibrium with each other. Within such a *biphasic* region, any variation in overall composition of the system will change the

Fig. 4.3 Plots of free energies of the minima for both anisotropic and isotropic phases on Fig. 4.2 as a function of composition expressed as volume fraction polymer, v_{p}. The free energy is plotted as $\Delta G/RT = (\ln Z - \ln Z_{(\bar{y}=1)})$. ($x = 75$, $\chi = 0$). The dashed line shows the common tangent construction, and the anisotropic line begins at v_{p}^{*}.

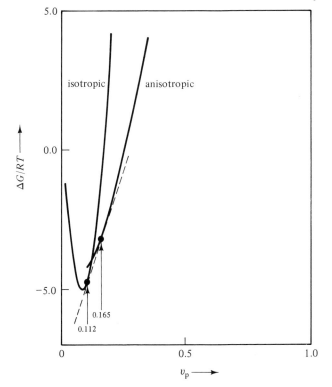

relative amounts of the coexisting phases but not their individual compositions. For two phases to coexist in equilibrium, it is necessary that they have the same chemical potentials, which is equivalent to saying that their partial derivatives of free energy with respect to n_s and n_p, are equal. This thermodynamic statement is embodied as the *common tangent construction* shown on Fig. 4.3. The composition range between the two tangent points, i.e. from $v_p = 0.112$ to $v_p = 0.165$, corresponds to the biphasic region. Below a concentration of 0.112, only the disordered phase is stable while above 0.165, there is again a single phase, but this time it is anisotropic. Figure 4.4 shows how the composition of the two coexisting phases, and thus the range of the biphasic region, is predicted to depend on the axial ratio x, calculated from the common tangent construction of the 1956 Flory model. It is interesting to note that the biphasic regime 'bridges' the metastable region just above v_p^*, discussed previously, so problems of

Fig. 4.4 Phase diagram showing the calculated composition ranges for the isotropic and anisotropic phases, and also the biphasic region in between, as a function of the axial ratio of the rods, x.

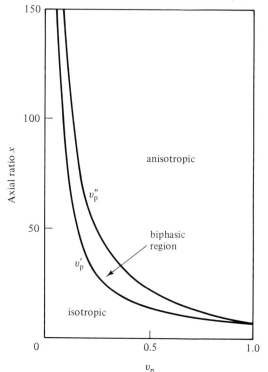

metastability are not significant. The concentration of the isotropic phase in the biphasic region is represented by v_p', and that of the anisotropic phase by v_p''.

The phase diagrams calculated above are invariant with temperature. However, the assumption that the solvent used is *athermal* ($\chi = 0$) is far from realistic as is shown by the temperature–composition phase diagrams already discussed (Chapter 3, e.g. Fig. 3.12). In mapping out a full phase diagram the dependence on χ (which will be a function of temperature) must be examined. However, it should be noted at the outset that χ is not a 'molecular shape' term and thus from this point onwards the model is no longer a purely steric one. The introduction of a finite χ leads to the addition of a term $\chi x n_p v_s$ to equation (4.14) for $-\ln Z$, where v_s is the volume fraction of solvent. If χ is negative the net effect on the biphasic equilibrium is slight. A negative χ corresponds to a 'good' solvent, in which the enthalpy decrease contributes to the free

Fig. 4.5 Free energy–composition plot similar to Fig. 4.3 ($x = 75$), except it is calculated for a positive value of χ of 0·015. Note that the common tangent is now between the isotropic curve at low composition and the anisotropic curve at a polymer composition approaching unity.

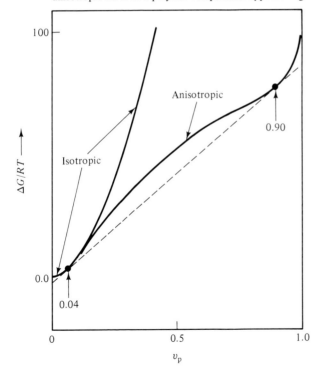

energy reduction on solution. However, if χ is positive there is a tendency for the polymer and solvent to segregate, and the effect on the extent of the biphasic region (i.e. the range of values of v_p for which isotropic and anisotropic phases can coexist) is dramatic. For a fairly small positive value of χ (~ 0.07, but dependent on x) the biphasic region suddenly increases in width as the concentration of the anisotropic component moves right across towards the right hand side of the diagram. The reason for this behaviour is apparent from the energy plot in Fig. 4.5 for $x = 75$ and $\chi = 0.15$. The effect of the positive $\chi x n_p v_s$ term is to cause the free energy curve for the anisotropic phase to bow upwards in the mid-range of composition so that the common tangent with the isotropic curve does not touch the anisotropic one until much higher concentrations. Thus only a small attractive interaction between the rods is sufficient to lead to the emergence of two phases of very different concentrations, the dilute one being isotropic, the concentrated, anisotropic.

The phase diagram predicted by the model for $x = 75$ is shown in Fig. 4.6. It is plotted as a function of minus χ since χ decreases with

Fig. 4.6 Calculated phase diagram showing the regions of phase stability as a function of concentration, v_p, and χ (plotted as minus χ upwards). LC′ and LC″ define the two liquid crystalline phases which can coexist under the 'dome'.

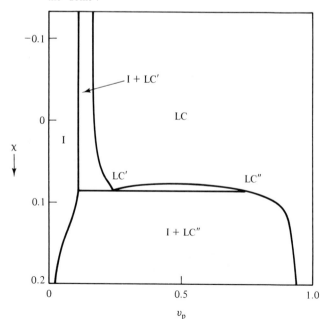

increase in temperature, and experimental diagrams are conventionally drawn with the temperature axis upwards. On the left hand side of the diagram (low v_p) all solutions are isotropic. At relatively small positive or for all negative χ's (high temperatures) the biphasic 'chimney' is entered as v_p increases. This chimney covers only a narrow concentration range before a single anisotropic phase forms. For more positive values of χ, or equivalently lower temperatures, the biphasic chimney is replaced by the much more extensive two-phase region. In contrast to solutions that fall within the chimney (where the difference in composition between the isotropic and anisotropic conjugate phases is slight) the phases in the broad biphasic region have very different compositions: the anisotropic phase possessing a v_p typically of ~ 0.8. Because of the slight upwards curvature of the nearly horizontal phase boundary separating the two-phase region from the single anisotropic phase region, the possibility exists of phase separation into two *anisotropic* phases over a narrow range of temperature or interaction parameter. Specifically, for $x = 75$, two anisotropic phases may coexist for $0.074 < \chi < 0.084$ (approx.). A triple point can therefore be identified at the cusp which is the intersection of the biphasic chimney with the flatter portion of the phase boundary. At this point two anisotropic and one isotropic phases coexist.

Before seeing how successfully the theory can describe real systems, we need to relate χ to temperature. The change in free energy of a polymer–solvent system is described by:

$$\Delta G = \Delta H - RT \Delta \ln Z \qquad (4.16)$$

However, the parameter χ is defined as $\Delta H / (x n_p v_s RT)$ so that we can write:

$$\Delta G = -RT(\chi x n_p v_s + \Delta \ln Z) \qquad (4.17)$$

It follows that, in order to predict a temperature–composition phase diagram with the broad two-phase region at lower temperatures and the narrow biphasic chimney at higher, the heat of mixing, ΔH, must be positive. At infinitely high temperatures the phase diagram would correspond to the $\chi = 0$ section of Fig. 4.6, but as the temperature is reduced χ will increase, and the diagram will then correspond to the positive χ part of the figure (albeit with a non-linear relationship between the vertical scales). Figure 4.7(*a*) shows the predicted temperature–composition diagram derived from Fig. 4.6, for molecular rods of axial ratio 75 and a constant value of χT of 22 K (so for $\chi = 0.05$, $T = 440$ K etc.).

Fig. 4.7 (*a*) Calculated phase diagram, as for Fig. 4.6, but with the vertical minus χ axis converted to temperature, compared with, (*b*) the experimentally determined phase diagram for PBLG in dimethyl formamide (DMF). (After [4.9].)

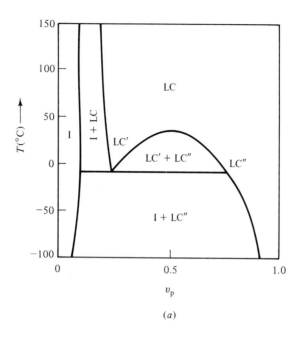

(*a*)

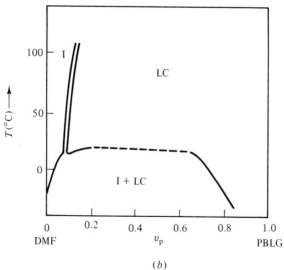

(*b*)

4.2.5 *Comparison with experiment*

The primary success of Flory's lattice model of liquid crystalline–solvent systems is the prediction of the main features of phase diagrams determined by experiment. In fact 'predict' is an appropriate word as the main theory was worked out before the observations were made. The model is elegant in that it requires knowledge only of the axial ratio of the molecular rods and a reasonable estimate of the heat of mixing of the solvent and the polymer. Comparison between the experimental diagram shown in Fig. 4.7(*b*), and the predicted diagram of Fig. 4.7(*a*) shows that the general form is correct, and suggests that a quite respectable fit could be obtained through the choice of appropriate values for χ and the axial ratio, x.

It is important to emphasize that neither the model nor the particular set of PBLG data chosen for comparison shows any effects associated with crystallization or gelation at high polymer concentrations. In most experimental systems, these factors will either totally change the form

Fig. 4.8 Plot showing the calculated concentration limits of the narrow biphasic region, v'_p and v''_p, (continuous lines) as a function of axial ratio, x, (as Fig. 4.4) compared with data on PBLG molecules of a range of molecular weights and in a variety of solvents. (Data collected in [4.2] and [3.27], the dashed lines were fitted to the points.) PBLG in: ($\bigcirc$, $\bullet$), dioxane; ($\square$, $\blacksquare$), dimethyl formamide–methanol; and ($\triangle$) *m*-cresole. The hollow symbols represent v'_p, the filled ones v''_p.

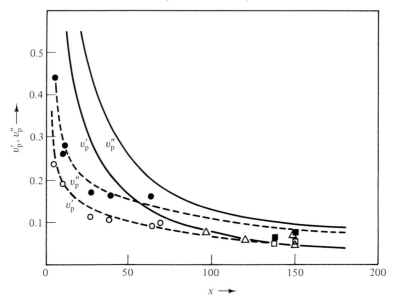

of the high concentration part of the diagram as shown in Fig. 3.17, or at least affect the kinetics and thus interfere with its precise determination.

A quantitative test of the theory is provided by its predictions of the limiting compositions v'_p and v''_p of the biphasic chimney, for $\chi = 0$. Figure 4.8 shows calculated values of v'_p and v''_p plotted as continuous curves, compared with data for samples of poly(γ-benzyl-L-glutamate) (PBLG) $\langle 22 \rangle$, of various molecular weights (and hence axial ratios) in a variety of solvents. If we bear in mind the simplicity of the Flory lattice model, the level of agreement between theory and experiment is quite encouraging. The model is, in essence, correct for molecular weights typical of polymers and provides a good foundation both for refinement and adaptation to other situations such as where molecules are not completely rigid. The poor agreement at low molecular weights may reflect the influence of long range interactions between the molecules as discussed in Section 4.3 below.

4.2.6   *The effect of polydispersity*

The original Flory formulation is for a system of identical rigid rods. In real systems, there will normally be a range of molecular lengths, i.e. the system is *polydisperse*. The lattice theory can readily be extended to cover this situation. It is found that all rods possess the same average $\bar{y}$, except those which are so short that $x < \bar{y}$. However, the ensuing degree of disorientation given by $\sin \theta = \bar{y}/x$ will differ for different rods, because of its dependence on x.

Flory and coworkers have looked in detail at the implications of polydispersity on theoretical predictions [4.10, 4.11, 4.12]. By choosing model systems with a specified distribution of rod lengths, such as the 'most probable' distribution (a distribution determined by the polymerization reaction; the distribution is discussed in reference 4.13), it has been shown that significant *fractionation* occurs between the isotropic and anisotropic phases. In other words the longer rods (larger x) preferentially occupy the anisotropic phase, and the shorter, the isotropic. Figure 4.9 shows the results calculated for the distribution of the weight fraction W of rods, of different axial ratio x, between the anisotropic and isotropic phases.

A second predicted consequence of a range of axial ratios being present is an increase in the difference in composition between the coexisting isotropic and anisotropic phases over that for an ideal monodisperse system. Comparison with experiment shows qualitative, but not quantitative agreement. The ratio v''_p/v'_p is a convenient quantity to measure. For monodisperse rods, theory predicts this ratio to be

1.46 for the limiting case of $x \to \infty$. The experimental values for (polydisperse) PBLG and other polymers are rather greater [4.14, 4.15, 4.16], but most of the results show that v_p''/v_p' lies in the range 1.4 to 1.8, which is rather lower than the calculated value of 2–3 for a system of rods with the most probable distribution. Nevertheless, because of the difficulty in carrying out the experiments, this agreement must be considered fair.

Despite these quantitative discrepancies, the tendency towards fractionation is always manifest [4.17, 4.18]. However, the degree to which it occurs is less than predicted. To some extent this may be due to sluggish diffusion between the two phases, with a consequent deviation from the equilibrium compositions remaining at finite times.

4.3 Maier–Saupe theory and orientation dependent interactions

4.3.1 *The Maier–Saupe approach*

This theory, first introduced around 1960 by W. Maier and A. Saupe [4.5, 4.6], has been extensively applied to small-molecule liquid crystals. It is a 'mean field' theory in that the energy of a rod molecule does not depend on its *particular* environment in any specific way.

Fig. 4.9 Plot of weight fractions, W, of molecules with a 'most probable' distribution of rods in the isotropic (W') and anisotropic (W'') phases. The uppermost curve, W° represents the unseparated (total) distribution with a peak at $x \approx 20$. The peaks in the two fractionated distributions occur at an axial ratio of 10.4 for the isotropic and 36.5 for the anisotropic phase. [4.11]

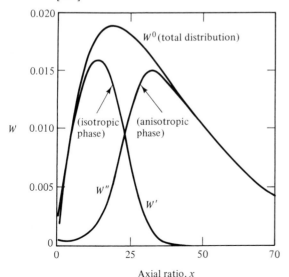

Axial ratio, x

Instead its energy is a function of its orientation with respect to the molecular director and of the degree of preferred orientation of all the molecules expressed as the *order parameter*, S. In describing the orientation distribution in terms of S (Chapter 2) the assumption is implicit that the distribution has a particular form, i.e. one which is describable with just the first order term of the even expansion of a harmonic series, for a general distribution would require several higher order terms as well. The free energy of the system depends also on the entropy, which is decreased as a result of the preferred orientation. The starting point for the derivation of the free energy is the orientation dependent part of the potential energy u of a given rod molecule, labelled by the subscript j. It can be written as:

$$u_j = -\mathscr{C}S\tfrac{1}{2}(3\cos^2\theta_j - 1) \tag{4.18}$$

where $\mathscr{C}$ is a constant which scales the magnitude of the interaction and is taken here to be independent of pressure and temperature although dependent on the volume of the molecule. θ_j is the angle between the axis of the cylindrical rod molecule and the director. Recalling that the orientation distribution can be described, at least in first order terms by:

$$S = \tfrac{1}{2}(3\langle\cos^2\theta\rangle - 1)$$

then the mean value of u_j is given by:

$$\langle u \rangle = -\mathscr{C}S\tfrac{1}{2}(\langle 3\cos^2\theta_j\rangle - 1)$$
$$= -\mathscr{C}S^2 \tag{4.19}$$

and hence per mole as:

$$U = -\tfrac{1}{2}N_A\mathscr{C}S^2 \tag{4.20}$$

N_A is Avogadro's number and the factor $\tfrac{1}{2}$ eliminates pairwise redundancy. As the level of orientation order increases, from the random case where $S = 0$ to perfect alignment with $S = 1$, the internal energy decreases ($\mathscr{C}$ is positive). However, this decrease in internal energy is opposed by an increase in the entropy term, $-T\Delta S$, for S decreases with increasing orientational order, S. The next stage is to determine the amount by which the entropy of the system is reduced as a result of the anisotropic angular distribution. The partition function for a single molecule can be written:

$$Z_j = \int_0^1 \exp[(-u_j/kT)\,\mathrm{d}(\cos\theta_j)] \tag{4.21}$$

substituting from equation (4.18) and separating the constant term it becomes:

$$Z_j = \exp-(\mathscr{C}S/2kT) + \int_0^1 \exp\left[\frac{3\mathscr{C}S\cos^2\theta_j}{2kT}\mathrm{d}(\cos\theta_j)\right] \quad (4.22)$$

The entropy per mole compared with the isotropic state is thus:

$$S = N_A k\left\{(\mathscr{C}S/2kT) - \log\int_0^1 \exp\left[\frac{3\mathscr{C}S\cos^2\theta_j}{2kT}\mathrm{d}(\cos\theta_j)\right]\right\} \quad (4.23)$$

If the assumption is made that the volume does not change significantly with S, then the change in free energy can be written by combining (4.20) and (4.23):

$$\Delta G = -N_A kT\left\{(\mathscr{C}S^2/2kT + \mathscr{C}S/2kT)\right.$$

$$-\log\int_0^1 \exp\left[\frac{3\mathscr{C}S\cos^2\theta_j}{2kT}\mathrm{d}(\cos\theta_j)\right]\Bigg\}$$

$$= -N_A kT\left\{\frac{\mathscr{C}S(S+1)}{2kT} - \log\int_0^1 \exp\left[\frac{3\mathscr{C}S\cos^2\theta_j}{2kT}\mathrm{d}(\cos\theta_j)\right]\right\} \quad (4.24)$$

Figure 4.10 shows plots of the reduced free energy, $\Delta G/N_A kT$, against the order parameter, S, for a number of different values of $\mathscr{C}/kT$. When this parameter is greater than a critical value, i.e. at low temperature and/or with a strong orientational dependent interaction, there is one minimum corresponding to a comparatively large order parameter (0.6 or more). This is the prediction of a stable nematic phase. Conversely, for the case of weak or non-existent interaction and/or high temperatures, the lowest free energy corresponds to $S = 0$, the isotropic phase. However, for a critical value of $\mathscr{C}/kT$ of 4.55, which for fixed $\mathscr{C}$ is the critical temperature, $T_{\mathrm{lc}\rightarrow\mathrm{i}}$, the ordered phase has the same free energy as the isotropic one. This condition thus corresponds to the liquid crystalline to isotropic transition, and also to a particular value of S of 0.43. The theory thus indicates that the perfection of orientational order in the liquid crystalline phase decreases with increasing temperature until it reaches 0.43 when the phase transforms to the isotropic state. It has been especially successful in predicting the behaviour of small molecule systems, and its formulation makes it particularly convenient for exploring the influence of external fields on phase stability.

The theory however is not readily extendable to handle polymeric systems or systems involving a solvent. It is not obvious how the interaction parameter, $\mathscr{C}$, should vary with the axial ratio of the molecule. Maier and Saupe made $\mathscr{C}$ dependent on the reciprocal of the square of the molar volume, although more recent arguments have favoured a simple reciprocal relationship. Intuitively one might expect $\mathscr{C}$ to increase with axial ratio, but whatever the correct value for long rods, the significant point is that the mean field approach does not explicitly include the axial ratio as a parameter. Furthermore, it is not clear how one should decide how $\mathscr{C}$ will change on dilution. It will of course decrease as the rod molecules move apart, but the exact function is not readily accessible. For these reasons the Maier–Saupe theory is not seen as being directly applicable to liquid crystalline polymer systems, although the idea of an orientationally dependent interaction energy is useful to polymeric theories, both in regard to the refinement of rigid rod models and the handling of partially flexible molecules.

Fig. 4.10 Plot of the reduced free energy, $\Delta G/N_\mathrm{A}kT$, against the order parameter S as predicted by the Maier–Saupe theory. The curves represent different values of temperature, while the one for which the two minima have equal free energy corresponds to $T_{\mathrm{lc}\rightarrow\mathrm{i}}$. [4.19]

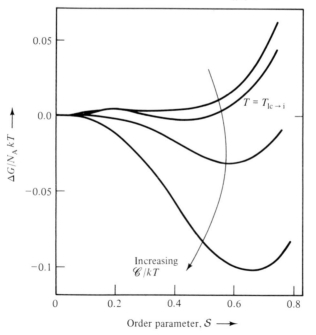

4.3.2   *Introduction of an orientation dependent interaction into
          the lattice model*

In the Flory lattice treatment, the enhancement of contacts between a pair of molecules as the orientational order increases is compensated by the corresponding elimination of contacts with other molecules, and for the case where there is no dependence of volume on the degree of order, the compensation is complete. However, rod molecules which have significantly anisotropic polarizabilities can possess orientationally dependent attractive interaction energies which will have an influence on the stability of the liquid crystalline phase additional to that of the molecular shape. This is particularly true for molecules including groups such as —N=N—, —C≡N and *p*-phenylene.

The orientation energy term of equation (4.20) is written to make it compatible with the lattice model as follows [4.8]:

$$E_{\text{orient}} = -\tfrac{1}{2}xn_p kT^*v_p S^2 \qquad (4.25)$$

In this relation the interaction energy ΔU is expressed by the characteristic temperature T^*, such that $\Delta U = kT^*$. The interaction energy is also proportional to r^{-6} and $(\Delta\alpha)^2$, where r is the mean distance between neighbouring segments and $\Delta\alpha$ is the anisotropy of polarizability. The introduction of v_p maintains the applicability to dilute systems as well, although the r^{-6} term would appear rapidly to

Fig. 4.11 The dependence of $T_{\text{lc}\rightarrow\text{i}}$ on the axial ratio, x. The temperature is plotted as the reciprocal of the reduced temperature $T_{\text{lc}\rightarrow\text{i}}/xT^*$. The figure illustrates the degree of anisotropic intermolecular interaction, described by the characteristic temperature, T^*, necessary to maintain liquid crystallinity up to $T_{\text{lc}\rightarrow\text{i}}$ in situations where the axial ratio is less than 6.4. [4.8]

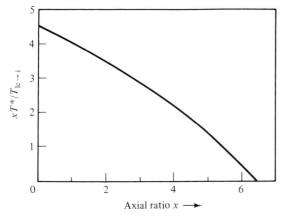

eliminate the orientation dependent term if the amount of solvent present became at all appreciable.

The orientation energy term plays a similar role in the free energy expression as the χ term did in the earlier development, although it is only applicable to the anisotropic phase for which a value of S must be assigned. Its presence means that thermotropic liquid crystalline phases can be stable at axial ratios less than 6.4 at sufficiently low temperatures. The effect is shown in Fig. 4.11, where the reciprocal of

Fig. 4.12 Phase diagrams calculated for rods of axial ratio 50 and 20, in a system in which $\chi = 0$ but there is an anisotropic intermolecular interaction. The vertical axis is xT^*/T (as with Fig. 4.11, except that here it is positive downwards), where T^* is a characteristic temperature dependent on the strength of the interaction. [4.20]

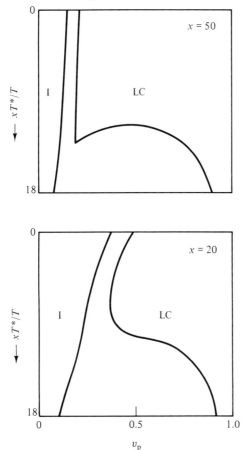

the reduced transition temperature $T_{lc \to i}/xT^*$ is plotted against the axial ratio.

It is also recognized that the introduction of the orientationally dependent term will influence the orientation distribution of molecular rods and thus the orientational partition function, Z_{orient}. Appropriate modifications have thus to be made in the exact derivation of this parameter.

Less attention has been paid to the case of lyotropic polymers, but it has been demonstrated that the presence of orientational dependent interactions without isotropic intermolecular interactions (i.e. $\chi = 0$) can lead to pronounced changes in phase stability with temperature and a phase diagram similar to systems with a positive χ [4.20]. For large axial ratios, the main distinction is that the biphasic chimney moves gradually to higher concentrations with increasing temperature but as the axial ratio decreases the differences become more marked. Figure 4.12 shows the predicted diagrams, plotted as a function of $-xT^*/T$, for $x = 50$ and $x = 20$. For the smaller axial ratio, the biphasic chimney widens and bends emphatically to the right while the cusp at the intersection of the broad and narrow biphasic regions is lost.

4.4 Polymers with semi-rigid chains

Up to this point the polymer molecule has been treated as an ideally rigid rod although in practice few macromolecules approach this ideal. DNA is one, while the tobacco mosaic virus, TMV, probably

Fig. 4.13 A smooth worm-like chain and an equivalent Kuhn chain consisting of rigid rods connected together by links which are completely free to move in all directions. The length of the rods is chosen so that the chain has the same physical characteristics, such as contour length and mean square end to end length, as the smooth one.

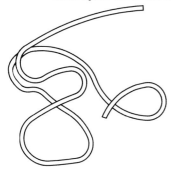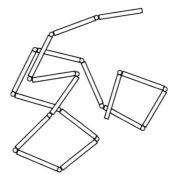

Worm-like (Kratky–Porod) chain   Kuhn chain

comes closest to a true rod-like form. One convenient way to describe less than completely rigid rods is in terms of a *worm-like chain* first discussed by Kratky and Porod [4.21]. For such chains each unit deviates modestly, but at random, from the direction of the preceding unit. As a consequence, chains comprising many units will possess significant curvature, and ultimately, as the number of units tends to infinity, begin to resemble a coiled chain. A useful way of treating a worm-like chain is to model it as a series of rigid segments connected together through completely flexible joints. Such an equivalent model is known as a *Kuhn chain*, and, although it is almost certainly a less accurate description of a macromolecule, it is more amenable to theoretical methods (Fig. 4.13).

4.4.1   *Critical persistence ratio*

The length of each segment of the Kuhn chain is equivalent to twice the *persistence length* of the molecule. This raises the possibility of whether the Flory lattice theory can be applied to worm-like molecules when treated as discrete rigid rods with a length equal to twice the persistence length of the worm. There is an initial indication that this may be a useful way of treating a semi-rigid polymer molecule when one considers that the rotation of a chain about its backbone

Fig. 4.14 Phase equilibria for the system; poly(*p*-phenyleneterephthalamide) (PPTA) ⟨44⟩–sulphuric acid. [4.22]

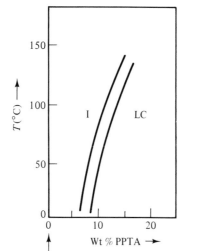

bonds, and thus its flexibility, is thermally activated. Hence, the hotter the molecule the shorter its persistence length and thus the shorter the equivalent rigid rods in the lattice model. It follows that the critical polymer concentration for liquid crystallinity in a lyotropic system, v'_p, would be expected to increase with increasing temperature and cause the biphasic chimney to curve across the phase diagram. Such behaviour is indeed observed, and a specific example is given for the system PPTA–H_2SO_4 in Fig. 4.14 (PPTA is the molecule $\langle 44 \rangle$).

It is possible within the framework of the lattice theory to calculate the effect of joining rods end to end through completely flexible links. The limitation affects only the combinatory contribution to the partition function, and has been shown to have a negligible influence on the predictions of the theory, at least for significant axial ratios [4.23]. Hence, the theory for rigid rods can be adapted to semi-rigid chains by the simple device of replacing the axial ratio of the rod molecule, x, by the axial ratio of the Kuhn segment, x_K, which is equivalent to the *persistence ratio* of the chain (persistence length/chain diameter). It should be emphasized that the x_K (persistence ratio) parameter used in the theory is a characteristic of the molecule in *isotropic* surroundings, and that it, and indeed the persistence length, would appear considerably larger if measured for the same molecule in the anisotropic phase.

In order to test the usefulness of the effective rigid-rod approach to polymer chains it is necessary to have some independent indication of the persistence length, and, through a knowledge of the molecular diameter, the persistence ratio. The persistence length can be measured in dilute solution with a variety of techniques such as light scattering, viscosity and flow birefringence. Taking hydroxypropyl cellulose ($\langle 21 \rangle$, from Table 3.3) as an example of a worm-like chain, its persistence length in dilute solution at room temperature comes out to be around 17 nm. This value leads to $x_K = 19$, and thus a predicted value of the critical concentration for liquid crystallinity, v'_p, of 0.38. Experimental observations for this threshold volume fraction in a variety of solvents give a mean value of $v'_p = 0.35$, which is quite reasonable agreement.

Measurements of persistence length are usually made in solution, which means that it is difficult to observe the influence of temperature over a substantial range. However, recent advances in computer based molecular modelling techniques enable predictions to be made of various chain parameters with increasing confidence. In this way it is possible to build molecular models with full chemical detail and also handle variants such as random copolymers. Furthermore, the models can be built at any temperature. As the parameters of interest apply to statistical chains, it is necessary to construct a sufficient number of

Fig. 4.15 (*a*) Computer generated models of chains of poly(hydroxy-naphthoic acid) ⟨14⟩ at 300 K and 800 K showing the statistical variation in their trajectories. (*b*) Plot of the mean persistence parameter of 100 chains of the same molecule at 300 K as a function of the number of units in the chain. The persistence parameter only becomes a non-extensive property of the polymer when the chain is long enough so that all orientational correlation is lost between the ends. The value characteristic of the infinite chain is given by the extrapolation of the plateau region of the plot onto the vertical axis. (Cerius software, CMD, Cambridge.)

(*a*)

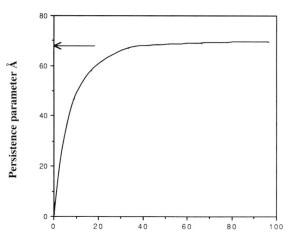

(*b*)   **Number of monomer units**

models in order to obtain a stable average value. Figure 4.15(*a*) shows a computer built chain of poly(hydroxynaphthoic acid) ⟨14⟩ at two temperatures. The first unit is positioned in the same orientation for each chain, and the deviations which follow for each model are associated with different rotations about the three bonds of the ester links. Figure 4.15(*b*) is a plot of the mean parameter of 100 chains measured in the direction of the first units against the total number of units comprising the chain. The length corresponding to the plateau of the curve is the persistence length of the infinite chain for each temperature. The great advantage of being able to assess the persistence length as a function of temperature is that it provides a means of predicting the liquid crystalline to isotropic transition temperature for molecules of different composition. Comparisons between predicted and experimental values for $T_{lc \to i}$ for a range of aromatic polyester molecules suggest that a critical persistence ratio of about 5 is appropriate. Figure 4.16 shows the quality of the prediction of $T_{lc \to i}$ for the copolymer based on hydroxybenzoic acid, isophthalic acid and hydroquinone ⟨34⟩. However, it is important to question the extent to which the persistence ratio criterion would be appropriate to a real chain which has the characteristics of a Kuhn chain, i.e. rigid sequences joined through flexible spacers as in molecule ⟨42⟩ (Table 3.6). In such a chain the critical ratio might be expected to be closer to 6.4 based on the axial ratio of the rigid units rather than on the persistence ratio where the persistence length is only $l_K/2$.

Fig. 4.16 Predicted values of $T_{lc \to i}$ for the system hydroxybenzoic acid–hydroquinone–isophthalic acid ⟨34⟩ using a critical persistence ratio of 5, compared with some experimental data (square points) (Fig. 3.6).

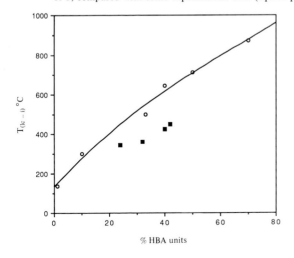

% HBA units

4.4.2   *Mean field theories for semi-rigid chains*

Another approach to modelling semi-rigid worm-like chains is to apply the mean field precepts of the Maier–Saupe theory. Two significant activities along these lines will be discussed. Ten Bosch and coworkers [4.24, 4.25] have, in essence, developed the Maier–Saupe theory to be applicable to semi-rigid chains, while Ronca and Yoon [4.26] have used mean field refinements in their extension of the original Flory lattice model to semi-rigid chains. Although attention will be focused here on these two pieces of work, parallel advances have been reported by others [4.27, 4.28, 4.29].

Ten Bosch *et al.* [4.24] considered a smoothly curving chain and defined its orientation at a distance n from one end (measured along the chain as the contour length) as a tangent vector, $\mathbf{r}_n$ (Fig. 4.17). All lengths are measured in units of the length of the monomer. The relationship for the partition function for such a chain can be shown to be:

$$\exp\left\{(-1/kT)\int_0^N [(\mathscr{B}/2)(d\mathbf{r}/dn)^2 + u_n]\,dn\right\} \tag{4.26}$$

In this expression the factor $\mathscr{B}$ is the bending constant which is related to the persistence length, q, by the relation:

$$q = \mathscr{B}/kT \tag{4.27}$$

The term $(\mathscr{B}/2)(d\mathbf{r}/dn)^2$ is then an expression for the elastic bending energy per unit length of chain as a function of the local curvature. The

Fig. 4.17 Description of a worm-like chain in terms of the local tangent vector.

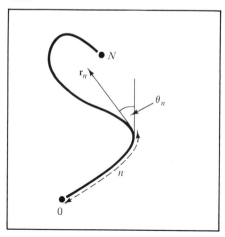

term u_n is the orientationally dependent mean field interaction and is the same term as that which forms the basis of the Maier–Saupe theory (equation 4.18), i.e.

$$u_n = -\mathscr{C}S_{\frac{1}{2}}^1(3\cos^2\theta_n - 1)$$

The integral in the statistical weight function is over the total contour length of the chain, N. The next step is to sum the individual statistical weights over all conformations of the chain. In order to include the influence of finite chain length this summation is carried out with the first and last elements set in fixed orientations, **R** and **R'**, to give the probability of a chain conformation as:

$$G(\mathbf{R}, \mathbf{R'}, N) = \int_{\substack{\text{All conformations,} \\ \text{with } \mathbf{R} \text{ and } \mathbf{R'} \text{ constant}}} \exp\left\{(-1/kT)\int_0^N [(\mathscr{B}/2)(\mathrm{d}\mathbf{r}/\mathrm{d}n)^2 + u_n]\,\mathrm{d}n\right\} \mathrm{d(conf)}$$

(4.28)

The overall partition function must now include the additional configurational possibilities associated with the range of possible orientations of the first and last elements. It appears as:

$$Z = \int_{\substack{\text{All orientations of} \\ \text{first and last elements}}} G(\mathbf{R}, \mathbf{R'}, N)\,\mathrm{d}\mathbf{R}\,\mathrm{d}\mathbf{R'}$$

(4.29)

The solution of this relation to give S as a function of $\mathscr{B}$, $\mathscr{C}$, and T is difficult. The route comes from the recognition that equation (4.28) satisfies a differential equation of the Schrödinger type which can be solved numerically. Further development of the theory is beyond the scope of this text, however some of the more important predictions are summarized in Fig. 4.18. The curves show that $T_{\text{lc}\to\text{i}}$ increases with increasing persistence length (it would also increase with increasing strength of the orientationally dependent interchain term, $\mathscr{C}$) and that at temperatures well below the transition the degree of order, S, is higher for the stiffer chain. However, for the range of parameters examined at least, the critical value of orientation, S^*, is constant at 0.35. The theory lies within the Maier–Saupe framework, although in addition to the anisotropic intermolecular term, there is also an *intra*-molecular one related to the chain stiffness and curvature.

The Ronca–Yoon model is based on the lattice theory in that the combinatory (self avoidance) component of the partition function remains an important factor. A mean field orientationally dependent energy term is introduced, as discussed in Section 4.3.2 above, which leads to an 'energetic' component of the partition function, Z_{en}. The rotational contribution to the partition function, Z_{rot}, is not applicable to semi-flexible polymers and is replaced by a conformational partition function, Z_{conf}, which has the form:

$$Z_{conf} = \int_{conformations} \exp\left\{-\mathscr{B}/2kT \int_0^N [(d\mathbf{r}/dn)^2 \, dn]\right\} d(conf)$$

(4.30)

The similarities with the bending energy function of the Ten Bosch theory (equation 4.26) are readily apparent. However, the conformational partition term is further modified to take account of the fact that the entropy of the chain is reduced by the additional orientation caused by the intermolecular energy term. A further refinement follows from consideration of the persistence length. For a constant value of $\mathscr{B}$ (the bending constant), q is inversely proportional to absolute temperature. It is clear however that mesogenic molecules are not smooth worm-like chains but are frequently built up from quite long rigid units which will set an effective lower limit to the persistence length. So, in addition to the persistence length, q, the model also includes a cut-off length, L_c, which is independent of temperature.

Fig. 4.18 Calculations of the order parameter, S, in the anisotropic phase as a function of temperature for chains with different persistence lengths, q. The chains are 100 units long, and the interchain interaction energy, $\mathscr{C}$, is 1.67 kJ mol^{-1}. (From Ten Bosch *et al.* [4.24].)

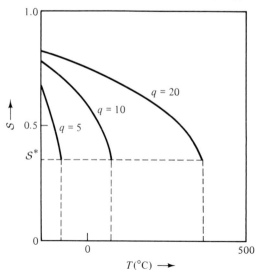

Figure 4.19 shows the dependence of $T_{\text{lc}\rightarrow\text{i}}$, normalized to $\hat{T}$ (the temperature at which the persistence length is reduced to the cut-off length), against the cut-off length expressed as the cut-off axial ratio x_c. Once the ratio reaches 4.45, the polymer is predicted to be liquid crystalline no matter how high the temperature. At lower values of x_c the transition temperature is lower, as liquid crystallinity requires the 'assistance' of an increasing persistence length and/or an orientationally dependent intermolecular energy. Figure 4.19 also plots the order parameter at the liquid crystalline to isotropic transition, S^*, against x_c. It is interesting that for $x_c = 0$, which corresponds to a chain which can become completely flexible at high temperatures, S^* is 0.44 which is close to the prediction of the Maier–Saupe theory and not very different from the critical value of 0.35 found by Ten Bosch *et al.* However, when x_c approaches 4.45, S^* increases to above 0.7 which is well above the range given by any simple mean field theory, and indeed closer to the prediction of 0.84 from the hard rod steric theory of Onsager [4.1]. The unmodified Flory steric model implies an even larger value [4.3].

4.4.3 *Chains with alternating rigid and flexible sections*

This class of semi-rigid chains is becoming of increasing interest as more and more thermotropic molecules are being synthesized

Fig. 4.19 Predictions of $T_{\text{lc}\rightarrow\text{i}}$ and S^* as a function of the 'cut-off axial ratio', x_c, (see text) from the theory of Ronca and Yoon [4.26].

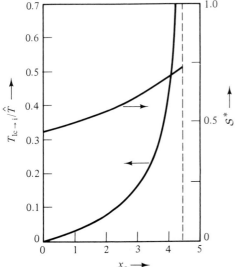

with this format. It is possible to treat such chains using lattice methods [4.30] on the assumption that the flexible spacers are sufficiently long to eliminate correlation between the successive rigid segments. The results of this approach indicate that as the disorientation, y, increases, the width of the biphasic gap decreases and likewise the extent of molecular weight fractionation for polydisperse species is decreased. However, implicit in the model is the assumption that the response of the flexible portions of the chain is independent of the overall orientation. While such a picture may emerge from a rigid-rod lattice model, it is now clear that if allowance is made for a generally orientationally dependent field arising from the liquid crystallinity (as in the mean field approach), then there will be significant stiffening of the flexible spacer [4.31]. Detailed analysis for a thermotropic system shows that, where the proportion of flexible spacer is low, the flexible units become oriented and contribute to the overall anisotropy of the mesogenic phase. On the other hand, a large proportion of flexible units will indeed diminish the degree of orientation and hence liquid crystalline stability.

4.4.4   *Helix–coil transitions and anisotropic phase formation*
In deriving the phase diagram of Fig. 4.7 it has been implicitly assumed that the molecule behaves as a rigid rod at all temperatures and in both phases with no conformational changes occurring. While the treatments of semi-rigid chains have taken into account the reduction of persistence length with increasing temperature, they have not envisaged much more dramatic conformational changes which can have a striking effect on the phase diagram. The 'helix–coil' transition is the prime example and is particularly important for biological macromolecules and their derivatives, synthetic or otherwise. PBLG has already been described as a rigid rod, made up of an α-helix stabilized by intramolecular hydrogen bonding. Many other polyamino acids and polypeptides also fit this description. However, a change in environment, notably pH or temperature, may destabilize the α-helix, by disrupting the hydrogen bonding, and the molecule then adopts a flexible coil conformation. Whereas, in the rod state (helical conformation) anisotropic phases may occur (they are often cholesteric because the biological molecules tend to be chiral), after the transition to a coil the possibility of anisotropy is lost. Thus the conformational change and the phase transition are likely to be strongly coupled.

Theoretical biologists have long been interested in the cooperative nature of the helix–coil transitions in proteins and related compounds [4.32]. For proteins many helical and coil-like segments may coexist in the same chain, but for homopeptides, such as PBLG, a transition from rigid to flexible can be identified over a relatively narrow temperature range (which is solvent dependent). Within this range the 'helical content' can be obtained by optical rotation measurements [4.33, 4.34].

The most detailed approach using the lattice model [4.35], introduces the statistical weights of the coil and helical sequences s_c and σ. In terms of these parameters, the equilibrium fraction of helical units is given by

$$\Theta = \tfrac{1}{2}\{1 + (s_c - 1)[(s_c - 1)^2 + 4\sigma s_c]^{-\frac{1}{2}}\} \tag{4.31}$$

From this starting point it is not only possible to show that the isotropic–anisotropic and coil–helix transitions are coupled, the latter becoming virtually discrete and approximating a first order transition, but also to examine the phase diagram as a function of Θ. Figure 4.20 shows the result of these calculations. As the helical content decreases, the biphasic gap widens. This behaviour is not unlike the case of alternating rigid–flexible segments described in Section 4.4.3. Qualitatively this is not surprising since a 'helix–coil' sequence can also be described in terms of a 'rigid–flexible' sequence. The most important conclusion of this model is the appearance of 'induced rigidity', i.e. intermolecular orientational order promoting intramolecular conformational changes. There are indications of the effect in the appearance of the helical structure at high concentrations and enhanced cooperativity of the transition [4.36, 4.37, 4.38]. Also supporting the idea of induced chain rigidity is the appearance of a re-entrant isotropic

Fig. 4.20 The effect of decreasing helical content on the nematic–isotropic equilibrium. Curves are calculated using $x = 100$, $m = 10$ (m is the number of helical units required to generate a helix of unit axial ratio), $\log \sigma = -4$ and $(a, a')\,\Theta = 1$, $(b, b')\,\Theta = 0.8$; $(c, c')\,\Theta = 0.5$ (terms defined in text). The primed parts of the curves denote the anisotropic phase. (After [4.35].)

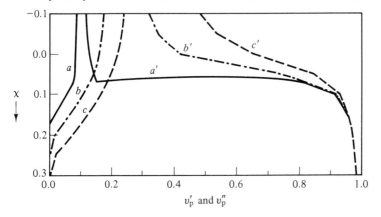

phase in PBLG in a 75:25% mixture of dichloracetic acid and dichloroethane as they are cooled [4.39]. (The phase is 're-entrant' in the sense that the phase diagram shows another isotropic phase for the same concentration solution at high temperatures.) The re-entrant isotropic phase occurs because the coiled chain is unable to support the liquid crystalline ordering, but the transition to the isotropic state occurs at a temperature lower than the helix–coil transition temperature measured in dilute solution.

4.5   Side-chain polymers

In Section 3.6.3 some reference was made to the possibility of new nematic phases occurring in side-chain polymers which contain also stiff elements in the backbone. These new phases arise because the backbone and side chain may either tend to lie parallel or perpendicular to each other. However, for side-chain polymers where the backbone possesses no mesogenic tendency, there is still a richness in the conformational possibilities, as has been explored by Warner and coworkers and reviewed in [4.40].

4.5.1   *Nematics*

Figure 3.21 (previous chapter) shows the three possible nematic phases suggested by Warner and Wang [4.41]. For flexible backbone side-chain polymers, the driving force to maximize the entropy of the backbone is antagonistic to the nematic tendencies of the side chain. If the side chains are to order as a nematic phase, then the backbone will deviate from the random walk conformation typical of a flexible chain polymer, and the molecular trajectory will tend to lie within an oblate spheroid. Neutron scattering experiments [4.42, 4.43], in which a small fraction of the chains are deuterium labelled to provide contrast in the melt, confirm this tendency.

4.5.2   *Smectics*

The mesogenic side chains of many liquid crystalline polymers will, by themselves, form smectic phases. It is not surprising that this tendency is carried over to the polymeric material. The organization of the side chains into layers means that the backbone itself is also similarly confined which significantly reduces its entropy. It has been proposed [4.44] that the backbone may sometimes jump from one layer to another, which will have a mitigating effect on the loss of entropy. The mechanism of such a jump may either involve a local loss of side chain order or the backbone moving through the mesogenic layers while making as small an angle as possible to them. The difficulty with the second route is that a large length of chain will reside in the high energy region. The actual trajectory will be a compromise between these two factors. Figure 4.21 illustrates the resultant walk of the polymer chain 'up and down' the layers.

Fig. 4.21 A one-dimensional random walk up and down the layers in a side chain smectic. The backbone is predominantly confined to a layer structure. (The side chains which form the layers are omitted for clarity.) (After [4.40].)

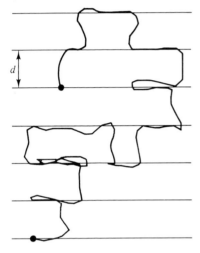

5   Local order and classification

5.1   Introduction

Friedel in 1922 [5.1] classified liquid crystalline phases as nematic, cholesteric and smectic, and the structural basis for each of these classes has been set out for small-molecule mesophases in Chapter 2. The nematic and cholesteric classes are characterized by intermolecular orientational correlations which are long range compared with molecular dimensions. The smectic class, which has many sub-divisions, allows for various types of *positional* correlation in addition to the long range *orientational* correlations which are the basis of all types of liquid crystallinity.

The orientational order in liquid crystalline polymers is sufficiently similar to that seen in small-molecule materials for the Friedelian classification to remain appropriate. However, there are several features peculiar to polymeric mesophases which raise new issues in the application of Friedel's scheme.

This chapter will consider the evidence for different types of liquid crystalline order in polymeric systems, and show how they can be classified within the conventional framework.

5.2   Nematic polymers
5.2.1   *Principles*

As with small-molecule liquid crystals the most simple polymeric mesophase is the *nematic* phase. The molecular chain axes mutually align in relation to a single director, but there is no long range positional order which might contribute to a crystal lattice.

The simple rod model of the nematic phase introduced for small molecules (Fig. 5.1(a)) can be extended to the polymeric state by using longer rods as shown in Fig. 5.1(b). The diagrams suggest that the distribution of orientations about the director will be more restricted in the polymer case so that the quality of the alignment will be better. More realistic models of mesogenic polymer molecules can help to explain both modifications to the stability of the nematic phase and the occurrence of other phases such as cholesteric and smectic. Polymer molecules which form mesophases are rarely completely rigid, although examples such as poly(*p*-phenylene) ⟨15⟩ and PPTA ⟨44⟩ may be nearly so. They are usually best thought of as *semi-rigid chains*, and theories incorporating molecules of this type have been described in the previous chapter. A chain that is too flexible will not form a liquid crystalline phase, but given sufficient rigidity the chains will align with each other as indicated in Fig. 5.1(c). Even though the chains wander to some extent in the mesophase, it should be emphasized that they will be straighter than would be expected for an isolated molecule, or in an isotropic phase under the same conditions. There are also theoretical [5.2] and experimental [5.3] indications that the quality of alignment in the mesophase, as defined by the *order parameter S*, will be a direct function of the stiffness. However, the important point to make at this stage is that although modifications of the simple rigid-rod model, to

Fig. 5.1 Schematic diagrams of nematic organization of molecules: (a) a nematic small-molecule liquid crystal; (b) a nematic liquid crystalline polymer with rigid chains; (c) a nematic liquid crystalline polymer with semi-rigid chains. The director, **n**, is vertical in each case.

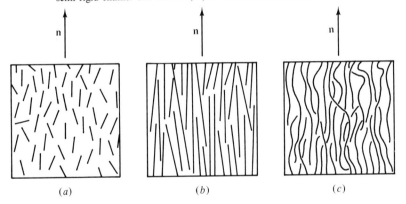

(a) (b) (c)

Fig. 5.2 A representation of local orientational order about all three axes, which would give rise to a biaxial nematic phase. It is important to appreciate that the laths could equally well represent fibrils (within which the mutual alignment of molecules is much better than it is between them), as it could ribbon-shaped molecules.

account for limited chain flexibility, may predict a reduction in liquid crystalline stability (as discussed in Chapter 4), such flexibility cannot explain the presence of any mesophase other than nematic.

Another type of nematic phase is the so-called *biaxial nematic*. In this case chain molecules with anisotropic cross sections are so organized in a mesophase that there is long range orientational order about all three orthogonal axes. In other words, not only do the molecules lie with their chain axes parallel, but there is also rotational correlation about these axes. This concept is illustrated for lath-like chains or fibrils in Fig. 5.2. The reason why the presence of full, three axis, rotational correlation leads to the term 'biaxial' needs some explanation. It stems from the description of the optical properties of a non-uniaxial system. There will be three principal refractive indices, but only two directions along which the optical behaviour will be isotropic. These directions are known as the *optic axes*. Thus, orientational correlation in three dimensions means that there are two optic axes, rather than one as in the case of a conventional uniaxial nematic, and this has led to the use of 'biaxial nematic' as a structural classification.

5.2.2   *Identification of nematic polymers*

In many respects the identification of a nematic state in a polymer uses the same methods developed for small-molecule meso-

Fig. 5.3 Threads characteristic of a nematic texture in polymer ⟨33⟩ seen in transmitted light without polars. (Courtesy Dr T. J. Lemmon.)

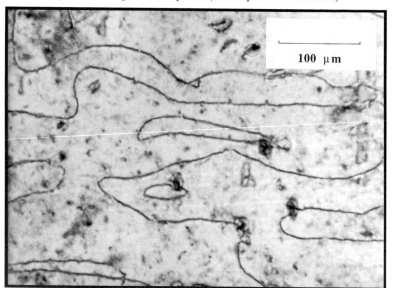

100 μm

phases. The name 'nematic' implies the presence of threads ($\nu\eta\mu\alpha$) seen in the transmitted light microscope without crossed polars. Figure 5.3 shows an example observed in a liquid crystalline copolyester $\langle 33 \rangle$. It is important to appreciate that such threads are made visible through phase contrast which, where specific microscopic attachments are not used, is derived from the aperture restriction in the back-focal plane.

It follows that the thread contrast will be more apparent with low magnification objectives which have a low numerical aperture. Figure

Fig. 5.4 A Schlieren texture typical of a nematic phase seen between crossed polars. Note twofold and fourfold brush arrangements such as those centred at 'A' and 'B' respectively. The polymer is again $\langle 33 \rangle$, and it had been quenched from the mesophase to room temperature. (Courtesy Dr T. J. Lemmon.)

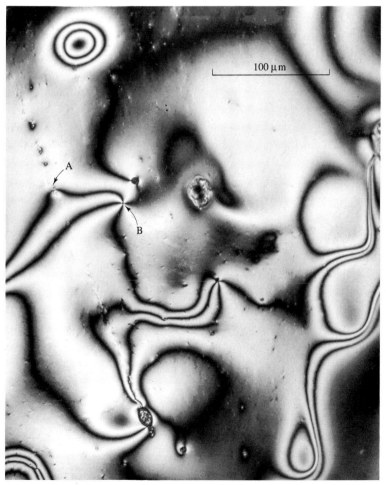

5.4 is an example of a Schlieren texture. It is another characteristic of the nematic phase, and is apparent when the sample is examined between crossed polars. The texture has also been described by Friedel as *structures à noyaux* (noyaux = nuclei). The dark regions occur where one of the principal axes describing the local optical properties, in the section plane, is parallel to either the polarizer or crossed analyser. Rotation of the crossed polars causes the dark regions to move to different regions of the sample, where the correspondence between the direction of the principal axes and the polars is again satisfied. The presence of point singularities which appear to be the centres of groups of either two or four radiating dark bands or 'brushes' can be taken as a clear indication of a nematic phase. (Examples of *both* two and four brush singularities must be present to confirm the nematic phase.) The brushes rotate when the crossed polars are rotated, rather like small propellors.

Another approach to phase identification in liquid crystalline polymers is to look for complete miscibility with an already identified small-molecule mesophase of similar chemical character. An example is shown in Fig. 5.5 where a basic phase diagram has been plotted for a liquid crystalline polymer ⟨49⟩ and a reference molecule ⟨50⟩ which is

Fig. 5.5 Phase diagram for a binary system of a polymer, ⟨49⟩, and a low molecular weight reference material, ⟨50⟩, which is known to be nematic. (Redrawn from [5.4].) N ≡ nematic, C_P ≡ crystalline polymer and C_R ≡ crystalline reference molecule.

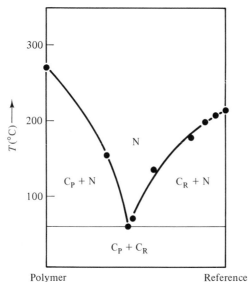

$$\left[-\bigcirc-CH=N-\bigcirc_{CH_3}-N=CH- \right]_n \qquad <49>$$

$$CH_3-O-\bigcirc-CH=N-\bigcirc-N=CH-\bigcirc-O-CH_3 \qquad <50>$$

known to be nematic. The fact that there is complete miscibility above the crystal melting point of the polymer at 265 °C, indicates that the polymer shows the same type of mesophase as the reference, in this case nematic. Such phase diagrams are usually prepared from a combination of observations of microstructures as a function of temperature and composition, and DSC thermograms. However, it should be emphasized that systems in which substantial chemical differences exist in the make-up of the polymer and the small-molecule reference are not likely to give miscibility over the complete composition range even though they may be isomorphous in terms of liquid crystalline classification. The method is also likely to be more difficult to apply to polymeric systems, where the kinetics of mixing may be sluggish. The technique can thus confirm but not rule out, the presence of a particular phase type in a mesomorphic polymer.

The X-ray diffraction pattern of an aligned nematic polymer will be similar to that shown in Fig. 2.5(*a*) and discussed in the section on diffraction in Chapter 2. The key features which characterize the nematic structure are the alignment with one symmetry axis, the presence of diffuse equatorial maxima and layer lines which are centred on the meridian and diffuse along their length. The existence of any sharp maxima is indicative of some type of long range positional order, such as occurs in smectic structures or crystals, and will take the polymer out of the nematic classification. However, nematic polymers may well be partially crystalline, and in that case sharp diffraction features will then be superimposed on a basic nematic pattern.

Biaxial nematics, in which there are correlations of molecular orientation about all three axes, are most readily identified from the biaxial symmetry of their optical properties. The standard method of deciding whether a crystal is optically uniaxial or biaxial is through observation of what are termed interference figures in the polarizing microscope using transmitted light. The figures are observed by illuminating the specimen with highly convergent light and inserting a so-called Bertrand lens below the eyepiece; this permits observations of the back-focal plane. Essentially the optical diffraction pattern of the sample is imaged, the interference figure being known as a conoscopic image. For a uniaxial material the figure consists of a dark cross coupled with concentric dark circles, its position depending on the crystal orientation. The cross will have four-fold symmetry; for some crystal orientations the cross point itself may be beyond the field of view. Figure 5.6(*a*) is an example of a uniaxial figure which is centred, and thus corresponds to the special case where the unique optic axis of the specimen is parallel to the microscope axis. For a biaxial sample the symmetry of the conoscopic image will be twofold. Figure 5.6(*b*) is such

Fig. 5.6 Optical interference figures of: (*a*) a uniaxial sample viewed along the optic axis; (*b*) a biaxial sample viewed along the bisector of the angle between the two optic axes. The cross polars are aligned orthogonally with the page as indicated by the arrows. (Redrawn from [5.5].)

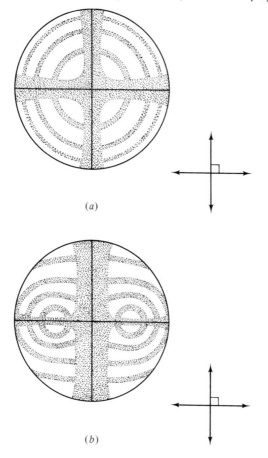

(*a*)

(*b*)

a biaxial figure. In this case the centring of the pattern implies that but, as in the uniaxial case (*a*), the specimen is in a special orientation in which the angular bisector of the two optic axes lies parallel to the microscope axis. The reader is referred to *Optical Crystallography* by E. E. Wahlstrom [5.5] or an equivalent text for a fuller account of the optical physics involved.

Main-chain thermotropic copolymers of the type developed for structural applications ⟨31, 32, 32, 33⟩ are generally considered to be nematic. However, they seldom show clearly defined optical textures, and these textures are frequently on such a fine scale that they tax the resolution limit of the optical microscope (of the order of 1 μm). A typical example is shown in Fig. 5.7. Under some circumstances these textures can appear as fine-scale versions of the Schlieren texture (Fig. 5.4), while the absence of obvious point singularities may be associated with the lack of sufficient resolution. The textures are reluctant to coarsen in the melt although they can be seen to be mobile at sufficiently high temperatures [5.7]. Furthermore, these polymers frequently show a lack of overall preferred orientation of the optically anisotropic regions, although the samples are manifestly anisotropic in the bulk, both in regard to structure (for example as shown by X-ray diffraction) and properties. These observations have been discussed in detail in [5.7] and can be rationalized if the polymer is assumed to be optically biaxial. Liquid crystalline polymers usually tend to fibrillate along the chain direction [5.8], but these polymers frequently show a

Fig. 5.7 Micrograph of a typical fine-scale texture of a random copolyester ⟨33⟩ as seen between crossed polars. [5.6]

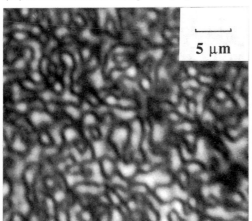

5 μm

preferred fracture plane after shear alignment, which again suggests some degree of biaxiality.

5.2.3   *The nematic phase in polymers with mesogenic side chains*

Although the nematic phase can be viewed as a liquid crystalline 'ground state' for both small-molecule liquid crystals and main-chain polymers, it is less common in polymers where the mesogenic groups are attached as side groups. The important role of flexible spacers, positioned between the polymer backbone and the mesogenic side groups, in enabling a liquid crystalline phase to form at all has been discussed in Section 3.6.1. Without spacers, the typical random trajectory of the backbone will hinder orientational alignment of the side groups. Alternatively if the side group order is able to force the backbone to adopt a straight conformation, the side groups will tend to be positionally as well as orientationally ordered and the structure will be smectic rather than nematic. Either way, a nematic phase is not likely to be observed.

The addition of flexible spacers of increasing length progressively decouples the mesogenic part of the side group from the backbone, and nematic order can be achieved. However, it is known from small-molecule systems that the smectic phase is encouraged when a flexible tail is attached to the mesogenic unit, where the tendency of the rigid and flexible parts of adjacent molecules to segregate opposite each other encourages the formation of layers. For the same reason, flexible spacers (which can each be thought of as a flexible tail attached to the mesogen rather than simply as a spacer) promote the formation of smectic phases in side-chain polymers. The nematic phase is normally observed in liquid crystalline side-chain polymers only where the length of the flexible spacer is between about two and six units. As one would expect, a flexible tail on the end of the side group removed from the backbone will also tend to destabilize the nematic phase with respect to the smectic. Many data are available to illustrate these trends, and some are quoted in Table 5.1 for molecules of the family ⟨51⟩ (similar to ⟨48⟩).

Nematic phases can be further stabilized with respect to smectics in side-chain materials by the substitution of side groups on the aromatic rings of the mesogenic unit; CH_3 is quite effective. Another way to achieve the same end is through random copolymerization which juxtaposes mesogenic side chains of different length, or alternatively, side chains with different lengths of flexible spacer.

It is not possible to picture the nematic phase of a side-chain mesogenic polymer completely, without wondering how the backbone

Table 5.1 *Transition temperatures of variants of a side-chain liquid crystalline polymer with a siloxane backbone* [5.9]

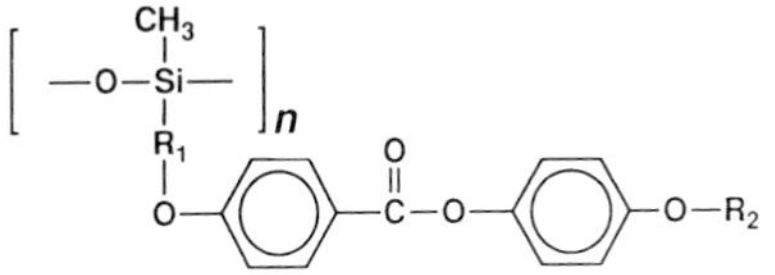

<51>

R_1	R_2	transitions (temperatures, °C):
$(CH_2)_3$	CH_3	glass $\xrightarrow{15°}$ nematic $\xrightarrow{61°}$ isotropic
$(CH_2)_6$	CH_3	glass $\xrightarrow{5°}$ smectic $\xrightarrow{46°}$ nematic $\xrightarrow{108°}$ isotropic
$(CH_2)_3$	C_6H_{13}	glass $\xrightarrow{15°}$ smectic $\xrightarrow{112°}$ isotropic

fits into the structure. One question which can be asked, and indeed answered experimentally using small angle neutron scattering, is what is the radius of gyration of the polymer molecules? This parameter is a measure of the envelope within which each molecule is positioned in space. Such measurements have been carried out on polymer ⟨52⟩ by Kirste and Ohm [5.10].

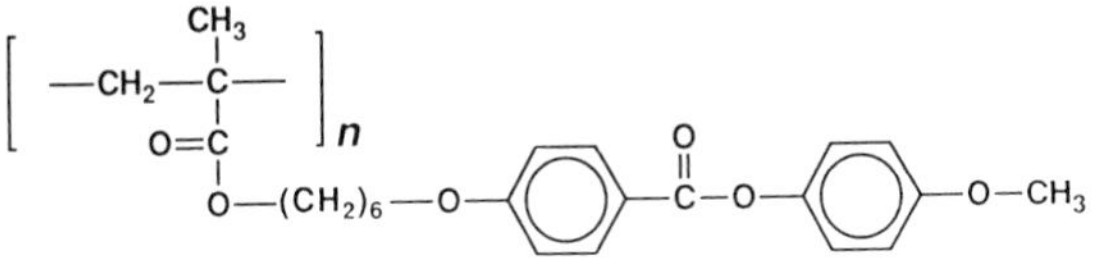

<52>

The polymer was mixed with material of similar molecular weight but with the backbone hydrogen atoms replaced by deuterium. The deuterated and hydrogenated chains will scatter neutrons differently, giving rise to contrast, and this enables the conformation of individual chains to be studied. Analysis of the small angle scattering showed that there is no detectable change in radius of gyration on passing from the nematic to isotropic states. The conclusion that can be drawn is that the liquid crystalline order of the side groups does not appear to have a significant effect on the trajectory of the backbone. However, if the side chains are aligned along the nematic director by a magnetic field, then the nematic order does cause the backbone to orient, on a modest scale, normal to the director. For example, measurements by these authors on samples of molecular weight of between 400000 and 450000 show that in the nematic phase the radius of gyration of the backbone is 9.4 nm parallel to the nematic director and 11.8 nm perpendicular to it. There is thus evidence, that despite the $(CH_2)_6$ flexible spacer, some measure of coupling remains between the side chains and the backbone.

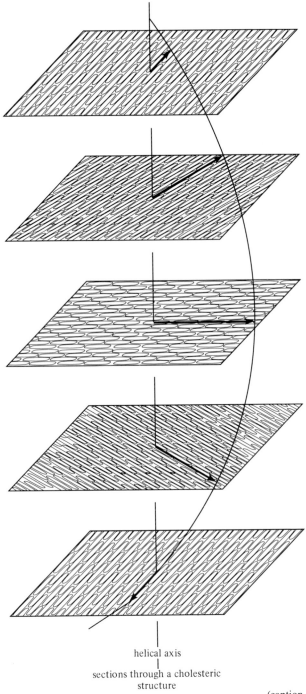

helical axis
sections through a cholesteric
structure

(caption opposite)

There have been some reports [e.g. 5.11] of optical biaxiality in nematic side-chain liquid crystalline polymers, based on analysis of interference figures seen normal to the plane of a thin specimen between glass slides. The interpretation of these observations in molecular terms must always be approached with caution, as the biaxiality may well arise from a preferred orientation of the backbones at some angle to the axis of the mesogenic groups (which might have been introduced for example by flow during sample preparation). This effect can confuse the main issue as to whether there is rotational correlation between the mesogenic units themselves.

5.3   Cholesteric polymers
5.3.1   *The cholesteric structure*

(The structure has been described in general for small-molecule materials in Chapter 2, Fig. 2.9(*b*).) On a very local scale, the arrangement of molecules in a cholesteric phase is not dissimilar to that in a nematic, and within an appropriately small volume the molecules can be ascribed a director. The difference between the two structures is that the cholesteric is twisted about an axis normal to the molecular director, which thus follows a helical path. However, this twist is gradual compared with the molecular dimensions. Figure 5.8 shows sections of a cholesteric structure for a small-molecule rigid-rod material. Only a few widely separated molecular layers are sampled in the diagram but the twist is quasi-continuous with a very small but constant angular increment for each molecular layer. The pitch of the helix in the diagram is about 100 molecular diameters (perhaps ~ 150 nm) which is small compared with many observed values. It is important to note that the twist in a cholesteric is spontaneous and the result of the *chiral* character of the molecules. It is also possible to twist a nematic artificially to give an equivalent structure, in which case it is referred to as *twisted nematic* rather than cholesteric. One method of doing this relies on the use of strong anchoring at the two surfaces, with the preferred alignment directions in the two surfaces not being parallel.

The pitch of the helix depends both on the temperature and, in the case of lyotropic systems, the polymer concentration in the solvent. Lyotropic solutions of PBLG ⟨22⟩ are cholesteric and Fig. 5.9 shows the influence of temperature on the helix pitch for this polymer in two

Fig. 5.8 The cholesteric structure in which the director twists comparatively gradually about an axis normal to the chain axes, to form a helical arrangement. (The degree of positional order in each layer, as illustrated, is larger than would occur in a mesophase.)

different solvents, the *reciprocal* of the pitch being plotted against temperature. It is interesting that in this example the sense of the helix rotation changes, and hence there is a specific temperature for each solvent for which the pitch is infinite. In other words the material becomes, transiently, a nematic, as the left and right handed tendencies compensate.

In a lyotropic system, the cholesteric twist tightens and the pitch shortens, as the polymer concentration increases. The exact power law depends on temperature but the behaviour can be represented by:

$$\mathscr{P} = \frac{1}{Ac^n} \tag{5.1}$$

where $\mathscr{P}$ is the pitch, A is a constant, c the concentration and n varies between 1 and 2. It has been suggested [5.13] that the combined effects of temperature and concentration on the helix pitch can be described by:

$$\mathscr{P} = \frac{1}{Ac^n (1 - T/T_n)} \tag{5.2}$$

where T_n is the compensation temperature (at which the polymer is nematic), and a negative pitch (when $T/T_n > 1$) implies a pitch of opposite sense.

Fig. 5.9 The temperature dependence of the pitch, $\mathscr{P}$, of cholesteric solutions of PBLG in (○) *m*-cresol and (●) benzyl alcohol. The pitch is plotted as its reciprocal. (Redrawn from [5.12].)

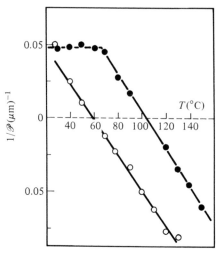

5.3.2   *Types of cholesteric polymer*

Cholesteric behaviour can be induced in an otherwise nematic polymer by the addition of small-molecule chiral compounds. Figure 5.10 shows a phase diagram prepared [5.3] for a system consisting of a nematic liquid crystalline copolyester $\langle 53 \rangle$ and a small-molecule chiral

$$\left[-\overset{O}{\underset{\parallel}{C}} - \bigcirc - O - \right]_{0.6} \qquad\qquad \langle 53 \rangle$$

$$\left[-\overset{O}{\underset{\parallel}{C}} - \bigcirc - O - (CH_2)_2 - O - \bigcirc - \overset{O}{\underset{\parallel}{C}} - O - (CH_2)_2 - O - \right]_{0.4}$$

Fig. 5.10 Phase diagram showing the cholesteric range of miscibility (Ch) between a small-molecule chiral mesogen $\langle 54 \rangle$, and an otherwise nematic random copolyester $\langle 53 \rangle$. Note the chiral smectic C, S_c^*, phase. [5.3]

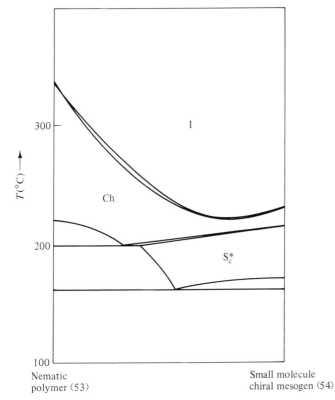

Nematic
polymer $\langle 53 \rangle$

Small molecule
chiral mesogen $\langle 54 \rangle$

$$C_4H_9-\overset{*}{\underset{\underset{CH_3}{|}}{C}H}-CH_2-O-\bigcirc-\bigcirc-\overset{\overset{O}{\parallel}}{C}-OH$$

⟨54⟩

∗   Chiral centre

mesogen ⟨54⟩. Very small amounts of the chiral component suffice to turn the nematic into a cholesteric, and in the appropriate temperature range a single cholesteric phase field extends across all compositions. Such behaviour is easily reconciled when a nematic is seen to be a special case of a cholesteric, but with infinite pitch. Cholesteric polymer

Fig. 5.11 Two approaches to the stabilization of a cholesteric phase at the expense of a smectic phase in a side-chain polymer. (*a*) Random arrangement of mesogenic chiral side groups with flexible spacers of different lengths. The effect has been demonstrated for the cholesteric ester side group ⟨55⟩ with spacer lengths of 2 and 12 —(CH$_2$)— groups distributed at random along a methacrylate backbone, the attachment being through a phenylene group [5.14]. (*b*) Random arrangement of chiral ⟨55⟩ and non-chiral ⟨56⟩ mesogenic side groups which effectively stabilizes the cholesteric rather than smectic phase when attached to a siloxane backbone. [5.15]

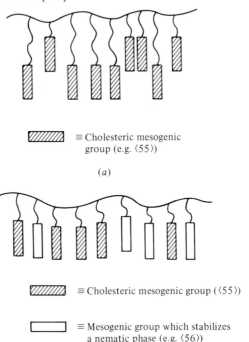

▨▨▨ ≡ Cholesteric mesogenic
        group (e.g. ⟨55⟩)

(*a*)

▨▨▨ ≡ Cholesteric mesogenic group (⟨55⟩)

▭▭▭ ≡ Mesogenic group which stabilizes
        a nematic phase (e.g. ⟨56⟩)

(*b*)

phases resulting from the addition of a low molecular weight chiral compound are known as *induced cholesterics*.

A more direct approach is to add the chiral element into the polymer chain itself, either by random copolymerization or by building a homopolymer directly from a monomer which is both mesogenic and chiral.

By far the most widely studied cholesteric polymer is the polypeptide PBLG ⟨22⟩, where the chirality of the molecule is implied by the 'L' for *Laevo*. In fact the external form of the molecule is a right handed screw and it is this which imparts the helicoidal twist to the structure. Mixed with its enantiomer, PBDG ('D' ≡ *Dextro*), in equal quantities, the resultant structure will be nematic rather than cholesteric. Many cellulose derivatives form lyotropic cholesteric phases, in suitable solvents, reflecting again the intrinsic chiralities of the natural polymer molecule. Some derivatives with relatively large substituted side groups such as hydroxypropyl cellulose ⟨21⟩ form thermotropic as well as lyotropic mesophases.

In the case of side-chain liquid crystalline polymers, the addition of side groups which are cholesteric as well as mesogenic onto a flexible backbone tends to introduce too much order with the result that a smectic rather than cholesteric phase is produced. Two modifications which destabilize smectic phases have been suggested by Finkelmann and coworkers [5.14, 5.15] and are illustrated in Fig. 5.11. In the first, the chiral groups ⟨55⟩ are attached to the backbone through flexible spacers of different lengths ($m = 2$, 6 and 12). This works well when the two lengths of flexible link are present in equal quantities. In a second approach, units with the same chiral mesogenic side group ⟨55⟩ ($m = 3$) are copolymerized with non-chiral mesogenic units such as ⟨56⟩.

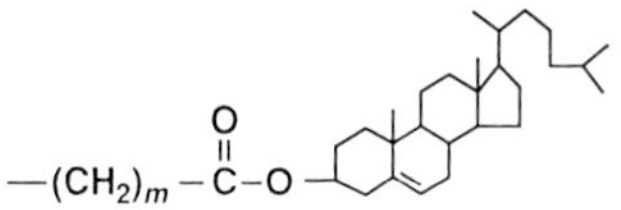

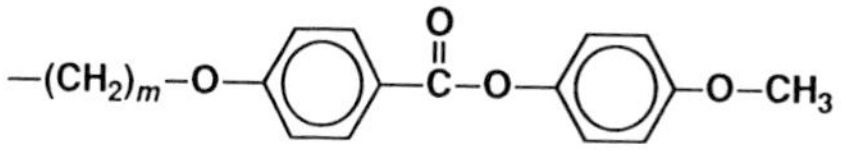

5.3.3 Optical properties of cholesterics

Many of the microscopic textures seen in cholesteric polymers can be directly related to the twisted molecular superstructure, as indeed can the unique optical properties of the bulk material. One of the most characteristic of cholesteric textures apparent in the polarizing light microscope consists of near-parallel sets of swirling lines. Not surprisingly they are known as *fingerprint textures*, and an example is shown in Fig. 5.12. The lines are associated with the layer-like optical periodicity along the helix axis, and their spacing is half the helical pitch. *Focal conic textures* are also seen, but unlike fingerprints, they are characteristic of smectic mesophases too. They are described more fully in Section 5.4.5 while their detailed explanation is deferred to Chapter 6.

Fingerprint textures occur when the helix axis tends to lie in the specimen plane. However, the proximity of boundaries, such as glass slides bounding a thin microscopic cell, can produce a texture in which the molecular axes are confined to the sample plane and the helix axis is normal to it. This geometry is sometimes referred to as the *Grandjean texture* and is represented in Fig. 5.13(*a*). Interestingly, if the confining glass slides are not parallel (5.13(*b*)), then a series of dark lines are seen in the microscope which correspond to the condition when the specimen thickness equals an integral number of semi-helix pitch translations. The lines are called *Grandjean lines* after Grandjean who first reported them in 1921 [5.17].

Fig. 5.12 Fingerprint texture in a lyotropic system of 20 % PBLG in dichloromethane. [5.16]. Molecular weight = 40 000.

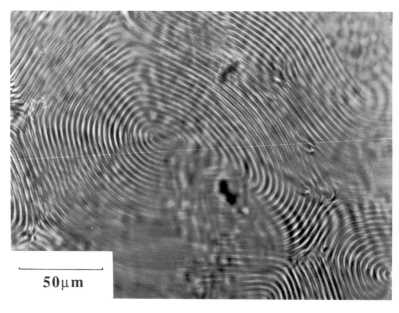

50μm

The special optical properties of cholesterics can be divided into four types:

(*1*) *Pseudo Bragg reflections.* Cholesteric polymer systems in which the helical pitch is of the order of the wavelength of light, show beautiful iridescent colours in white light. The colours vary with angle of incidence and observation, and change with temperature and polymer concentration in response to variations in the helix pitch.

Because, for a given orientation of the light beam, the refractive index of the sample will appear to vary periodically along the helix axis, light is diffracted when the Bragg type condition:

$$\lambda = n\mathscr{P} \sin \theta_{B} \tag{5.3}$$

Fig. 5.13 Representation of a Grandjean texture between (*a*) parallel slides and (*b*) non-parallel slides showing the condition for Grandjean lines where the spacing of the slides is an integral multiple of half the helix pitch. The nail convention is used here, where the length of the nail is proportional to the projected length of the molecule onto the page, and the head can be thought of as the end which points out of the page.

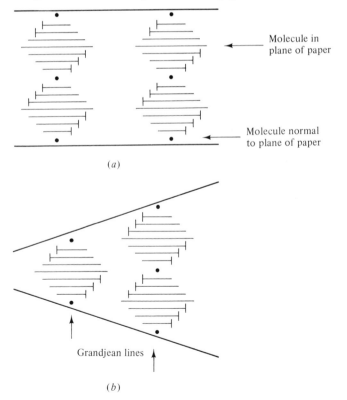

Molecule in plane of paper

Molecule normal to plane of paper

(*a*)

Grandjean lines

(*b*)

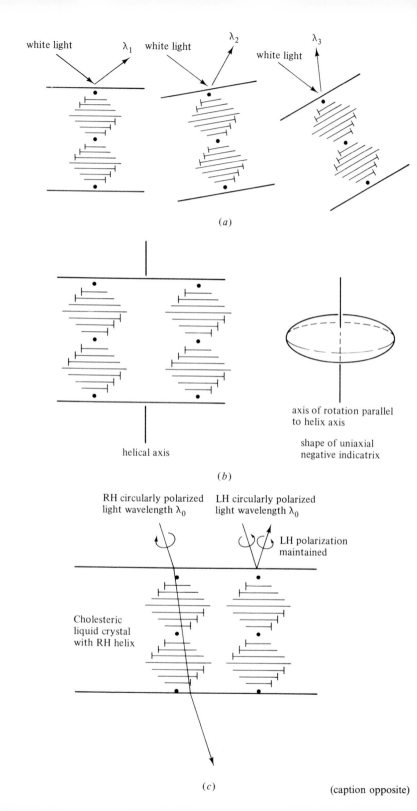

white light λ_1 white light λ_2 λ_3
white light

(a)

helical axis

axis of rotation parallel
to helix axis

shape of uniaxial
negative indicatrix

(b)

RH circularly polarized   LH circularly polarized
light wavelength λ_0 light wavelength λ_0

LH polarization
maintained

Cholesteric
liquid crystal
with RH helix

(c)

(caption opposite)

is satisfied, where n is the mean refractive index and θ_B the Bragg angle (Fig. 5.14(a)). As the periodic variation in refractive index will be sinusoidal for normal incidence, higher order diffraction effects are not observed in this case. For monochromatic light, the condition will only be satisfied for one particular angle, however with white light, differently oriented regions of a sample will select the wavelength appropriate to the Bragg condition from the incident light and diffract it, leading to multifarious coloured effects.

(2) *Optical symmetry axis.* In macroscopic terms (i.e. on a scale much larger than the pitch of the helix) an undistorted cholesteric structure is optically uniaxial, with the helix axis as the optic (symmetry) axis. As polymer molecules generally have a greater refractive index along the chain than at right angles to it, the refractive index along the helix axis will be lower than that perpendicular to it, and it is thus usual for cholesterics to be *optically negative* (Fig. 5.14(b)). If the pitch is persuaded to become infinite, either by adjusting the temperature or the concentration in a solvent, or by the application of an external field, then the polymer will become optically positive, and the optic axis, assuming the nematic phase is uniaxial, will switch through 90° to lie parallel with the molecular chains.

(3) *Circular dichroism.* Circularly polarized light can be thought of as being composed of two components, each plane polarized but with a $\pi/2$ phase shift between them. Depending on which plane of polarization is in advance, there are two senses of polarization (clockwise and anti-clockwise). In the case of a dilute isotropic solution of a chiral, and thus optically active molecule, circular dichroism occurs because one component of circularly polarized light is absorbed more strongly than the other. The twisted structure of the cholesteric mesophase produces circular dichroism, not through selective absorption of either the positively or negatively circularly polarized components, but through selective reflection. The effect, which is illustrated in Fig. 5.14(c), has its maximum strength at a particular wavelength, λ_0, where one component is almost completely reflected while the other is transmitted through the polymer. At normal incidence the reflective wavelength λ_0 is related to the helical pitch by the relation:

$$\lambda_0 = n\mathscr{P} \tag{5.4}$$

Fig. 5.14 Optical effects specific to cholesterics. (a) Pseudo Bragg reflections giving rise to a range of different colours when illuminated in white light. (b) Origin of optically negative behaviour of cholesterics. (c) Selective transmission and reflection for light at the 'reflective wavelength' λ_0, circularly polarized to the left and to the right.

where n is the mean refractive index. The reflected component does not have its sense of circular polarization reversed.

It has been the convention that a cholesteric liquid crystal which selectively *reflects* light circularly polarized in the left hand sense (anti-clockwise viewed in direction of propagation) is defined as left handed, and vice versa. Intriguingly, there can be some confusion here, as it is a mesophase with a *right handed* helix which selectively reflects the left handed circular component.

(4) *Rotatory power.* A typical optically active solution or crystal may rotate the plane of polarized light through anything up to 1–2 radians per mm pathlength. In contrast, plane polarized light transmitted through a cholesteric mesophase parallel to the helix axis can have its plane rotated by up to 100 radians per mm! These very high values of rotatory power are the consequence of the supermolecular organization in the cholesteric mesophase and do not directly stem from the character of the individual molecules. The rotatory power (Ω) can be estimated from the de Vries formula [5.18]:

$$\Omega = \frac{\pi \mathscr{P}}{4\lambda^2} \cdot \frac{\Delta n^2}{1-(\lambda/\lambda_0)^2} \tag{5.5}$$

where λ is the incident wavelength and (Δn) the birefringence in a plane perpendicular to the optical path and, in this example, to the helix axis as well.

Fig. 5.15 Schematic diagram showing the variation of optical rotatory power, Ω (solid curves), and circular dichroism (dotted curves) with wavelength. The reflective wavelength, λ_0, is equal to the product of the helix pitch and the mean refractive index of the polymer. (After [5.12].)

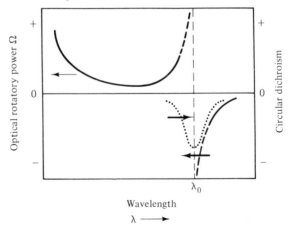

Unlike circular dichroism, the optical rotatory power is significant at wavelengths other than in the region of λ_0, although it does change sign at the wavelength. Schematic plots of optical rotatory power, calculated from equation (5.5), and circular dichroism as a function of wavelength are shown in Fig. 5.15.

5.4   Smectic polymers
5.4.1   *Smectic order and its identification*

While a nematic phase is characterized by long range *orientational* order but only short range positional order, a smectic phase shows, in addition, a degree of *long range positional* order which gives rise to a layer structure. An example of smectic organization in a small-molecule liquid crystal (described in Chapter 2), is sketched again in Fig. 5.16(*a*). At first sight it might seem that in smectic *polymers*, the chains could simply be lined up alongside each other, in the same way as small molecules, although the lamellae would be much thicker. This does not generally occur. In the first place polymers, unlike small mesogenic molecules or oligomers, have a distribution of molecular lengths so that any exact segregation of the ends of rigid chains into sharply defined layers is impossible. Secondly, as most liquid crystalline polymer molecules are not rigid rods on the scale of their length, the distance between ends is likely to be shorter than their maximum length and variable. For both these reasons, smectic polymers are unlikely to be arranged simply as thick versions of Fig. 5.16(*a*). Smectic phases occur in polymers because of molecular details which are not apparent in a smooth worm-like model of a simple chain.

Main-chain polymers in which rigid mesogenic sequences are separated by flexible sequences can show smectic behaviour: the two different types of sequence segregating so that they form layers normal to the chains. An example is shown in Fig. 5.16(*b*). In some instances the rigid sequences may induce orientational order into the flexible sequences as in the nematic phase of similar polymers, but such detail is not implied in the diagram. Although examples of such smecticity tend to be of the rigid sequence, flexible sequence type, there is no reason why this need be so. The chemically contrasting sequences could both be rigid, and the sequences could be as short as individual units.

Polymer molecules with mesogenic side groups attached to the backbone through a flexible spacer, often show smectic phases. In principle, the scheme is similar to that of Fig. 5.16(*b*). However, the backbone is likely to be accommodated in the regions between the mesogenic layers as shown in Fig. 5.16(*c*), although there is a possibility that parts of it may be incorporated within the mesogenic layers themselves.

Smectic type order leads to periodic variations in electron density in a direction normal to layers. The most appropriate technique for detecting the smectic phase is X-ray diffraction, where diffraction peaks, often at fairly low angles, correspond to the period of the density

Fig. 5.16 Examples of smectic type organization. (*a*) Small-molecule smectic. (*b*) Mesogenic polymer with flexible sequences in the backbone. (*c*) Polymer with mesogenic side groups attached to a flexible backbone through flexible spacers.

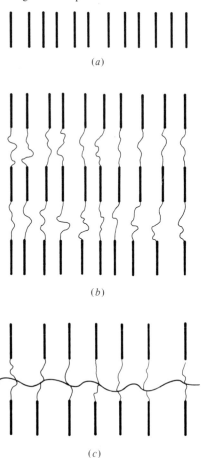

(*a*)

(*b*)

(*c*)

fluctuations. The forms of the diffraction patterns corresponding to different types of smectic polymers are shown in Fig. 5.17 (after [5.19]). Figure 5.17(a) is a representation of an X-ray pattern of the side-chain polymer ⟨57⟩. The maxima on the vertical axis indicate a layer periodicity of a few nanometers, whereas the diffuse maxima on the horizontal axis correspond to the spacing between the mesogenic side

Fig. 5.17 Diffraction patterns of smectic, side-chain liquid crystalline polymers. In each case, the fact that the wide angle reflections lie on the horizontal axis indicates that the mesogenic side groups, which make up the larger part of each molecule, are aligned vertically. (a) *Smectic A*. The low angle reflections lie on the vertical axis showing that the layers are horizontal. (b) *Smectic C*. The low angle reflections are disposed along axes on either side of the vertical which indicates that the layers are tilted. However, the inter side-chain reflections remain on the equator which means that the mesogenic sections of the chain are vertical and thus no longer normal to the layers. (c) *Smectic B*. The low angle pattern is similar to Smectic A, but the sharper wide angle maxima show that the mesogenic units are organized on a crystal lattice within the layers. (Representation of data in [5.19].)

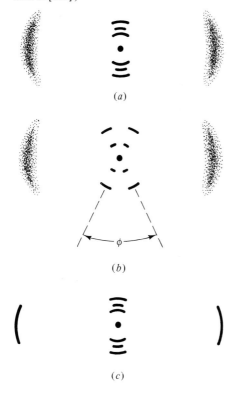

(a)

(b)

(c)

groups within the layers. Such a structure would correspond to the *smectic A* classification (cf. Chapter 2). Figure 5.17(*b*) shows smectic layer reflections at small angles which are arranged as an 'X' with an angle of ϕ between the arms. In this case the layers are tilted so that they are no longer perpendicular to the rod-like side groups within them. Such a pattern, which is from the side-chain polymer ⟨47⟩ (*m* = 11), is typical of *smectic C*. The *smectic B* phase in which the rods in each layer are arranged on a crystal lattice has also been observed in polymers. Figure 5.17(*c*) shows the X-ray pattern for this phase in a

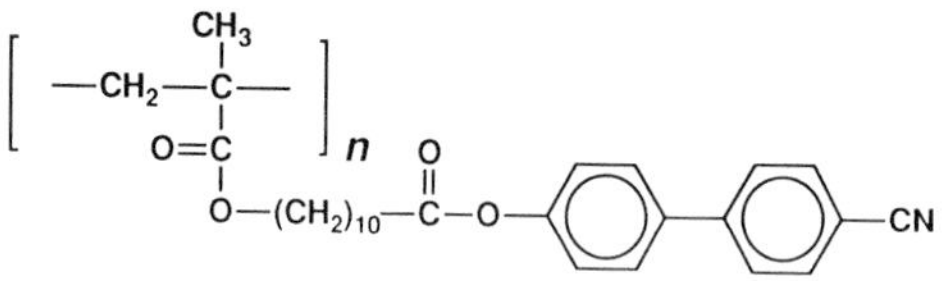

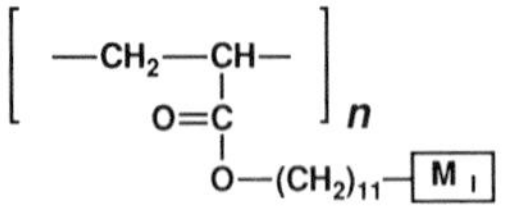

side-chain polymer ⟨58⟩. In this case the inter side-chain peaks on the horizontal axis are much sharper, and clearly indicate the long range positional order of the groups in the two-dimensional layers. It should be noted, however, that despite the additional order within the layers, the heat of transition to the isotropic melt is not much greater than for smectic A.

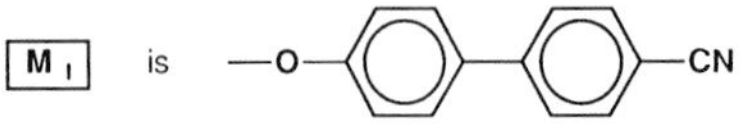

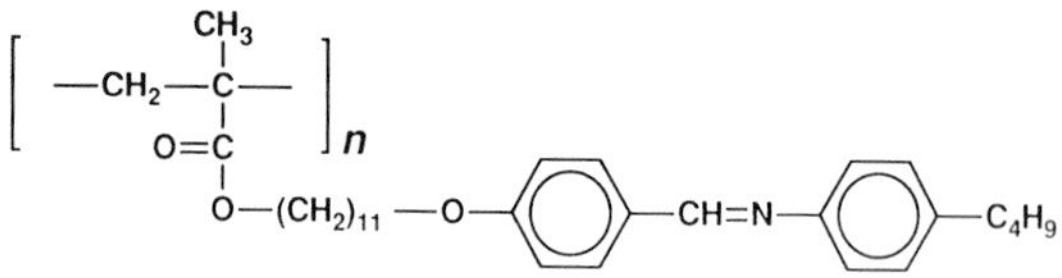

Measurement of diffraction patterns such as these, combined with a knowledge of molecular dimensions, can enable one to make fairly reliable models of the local structure, and some of these are now considered.

5.4.2 *Types of positional order in smectics*

The relationship between the layer spacing in a smectic polymer, as measured from X-ray diffraction patterns, and the length of the side group is not necessarily straightforward. Figure 5.18 shows situations in which the side groups interdigitate (*a*), abut (*b*) and overlap for the length of the flexible tails only (*c*). Only in (*a*) can the layer period be expected to correspond directly to the side-group length, and even then there is the backbone thickness to be accounted for.

Fig. 5.18 Packing schemes of mesogenic side groups which lead to different relationships between the layer spacings and the length of the side groups, after Shibaev and Platé [5.19]. (*a*) Interdigitating; (*b*) abutting; (*c*) overlap of flexible tails.

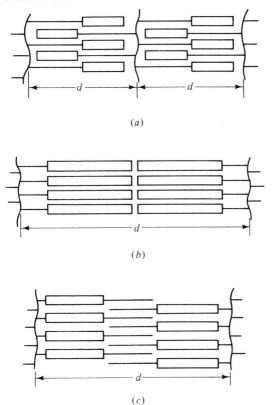

(*a*)

(*b*)

(*c*)

At this stage we need to reopen the question as to the role played by the backbone in smectic polymers. The presence of ordered side chains can increase the stiffness of the chain to the point where it is more sensible to view the backbone plus side groups as forming a ribbon-like chain. Such an approach has been followed by Zugenmaier and Mugge [5.20] for side-chain polymers with siloxane backbones, specifically for the molecule ⟨48⟩ with m equal to either 5 or 6. Figure 5.19(a) shows the molecule viewed as a ribbon with a near-rectangular cross section, while Fig. 5.19(b) shows the packing of the ribbons. Note that the backbones lie on a vertical set of evenly spaced lines, and would thus give rise to a sharp diffraction maximum characteristic of the smectic

Fig. 5.19 A smectic packing scheme for the side-chain polymer with a siloxane backbone ⟨48⟩ with $m = 5$ or 6. Note that it is the molecules which interlock, rather than the side chains which interdigitate. [5.20]

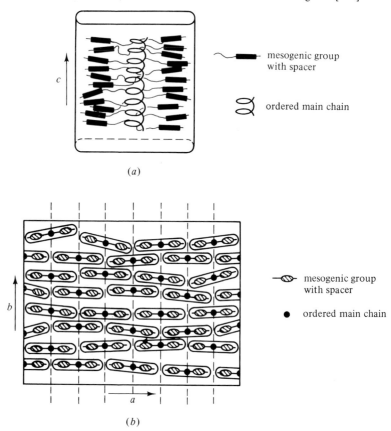

(a)

(b)

layers. It would be difficult to distinguish between the structure of
Fig. 5.19(*b*) and the interdigitated arrangement of 5.18(*a*).

If we relate the smectic layers to the ribbon shaped molecules, then
the chains lie *within* the layers. It is as if the basic organization is that
of a biaxial nematic phase, which becomes smectic on developing long
range positional order in either of the two dimensions normal to the
chain axes. Furthermore, it is not difficult to envisage a class of
mesogenic polymers which show long range positional order in two
dimensions. They fall short of being crystals, and yet will 'stretch' the
smectic classification as they are not really layer structures. The
simplest example of two-dimensional positional order in polymers is
that where smooth, straight rods are positioned on a two-dimensional
crystal lattice (strictly a 'net') with their axes aligned in the third
dimension. The rods are thus packed in a perfectly ordered array
without any longitudinal register between them. Such an arrangement
is seen in PTFE ⟨19⟩ above 37 °C, and in fibres of PBTZ ⟨45⟩ where
the packing is not hexagonal.

Viewing molecules as ribbons is also helpful in understanding the
smectic organization of the so-called 'hairy rod' molecules ⟨30⟩. In
this case the one-dimensional positional order is in a direction normal

Fig. 5.20 Measured layer spacing for a polymer with an aromatic polyester
backbone and aliphatic side chains ⟨30⟩ plotted against the number of
—(CH$_2$)— groups in the side chains, *m*. [5.21]

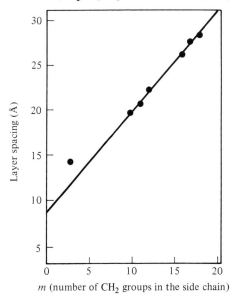

m (number of CH$_2$ groups in the side chain)

to the chain axis but in the plane of the ribbons. Figure 5.20 is a plot of the layer spacing determined from the position of sharp, 'smectic', diffraction peak against the number of $(CH)_2$ units in the side chain for a series of these molecules. The slope of the plot indicates that the spacing increases by 0.12 nm per additional $(CH)_2$ in each side chain, which corresponds quite closely to the 0.125 nm chain repeat for alkane chains in the extended conformation. Detailed arguments [5.21] suggest that the side chains of neighbouring molecules interdigitate and that the aromatic backbones are positioned in pairs with the side chains of each pointing in opposite directions.

The alternative arrangement, in which the long range positional order is in a direction normal to the planes of the ribbons (an arrangement referred to as a 'palisade structure') has been described for molecule $\langle 59 \rangle$ by Duran *et al.* [5.22]. Another unusual aspect of this particular side-chain polymer is that it apparently shows a liquid crystalline phase even though there is no flexible spacer between the aromatic part of the side chain and the backbone.

$$\left[-CH_2 - \overset{\overset{\displaystyle CH_3}{|}}{\underset{\underset{\displaystyle O}{|}}{\underset{O=C}{C}}} - \right]_n \qquad \langle 59 \rangle$$

Ribbon-like molecules which are sufficiently rigid to give liquid crystalline phases are perhaps better described as 'boards' or 'planks'. For this reason liquid crystalline phases in which the molecular organization reflects the anisotropic shape of the molecular cross section have been classified as *sanidic* ($\sigma\alpha\nu\iota\varsigma$ = board) [5.23, 5.24]. Without any level of smectic order, a sanidic phase is a biaxial nematic [5.25], but the classification would also include the smectic phases described above in which there is long range order in either of the two directions normal to the chain axis. A 'sanidic' class would thus overlap with established terminology, and the word might be better reserved as a description of molecular shape.

One other caveat is needed concerning these additional classifications of smectic order in liquid crystalline polymers. For, while a nematic polymer phase will tend to have a lower viscosity than the equivalent isotropic phase (Chapter 7), the development of smectic order tends to increase the viscosity to the point that the polymer will barely flow at all, especially if it is of high molecular weight. Furthermore, two-dimensional positional order will again increase the viscosity so that the polymer behaves much more like a solid than a liquid. Just to what extent such materials should still be considered *liquid* crystalline is a matter for debate.

5.4.3   *Poorly ordered smectics*

The segregation of rod-like molecules into layers is not in itself sufficient to ensure long range positional order in a direction normal to the layers. The layers themselves have to be periodically spaced on a one-dimensional lattice. However, when one is considering one-dimensional order there is no distinct phase change between long range order and short. Starting, for example, with a perfect, infinite lattice in one dimension, it is possible to introduce increasing amounts of paracrystalline disorder until the layers are spaced so irregularly that there is only short range order between them. The corresponding effect on the diffraction pattern will be that the sharp, smectic layer maxima will become diffuse, and the intensity of the higher order layer reflections much reduced as the maxima smear into each other. The result of the paracrystalline disorder is thus a broadening of the first order peak and the loss of the higher orders, as is illustrated in Fig. 5.21(*a*) and (*b*). Another type of disorder is associated with the loss of precision with which the molecules segregate into the layers which nevertheless retain their exact periodicity. In this case the density wave normal to the layers will tend to be sinusoidal with the result that the higher order diffraction maxima will be suppressed, even though the perfect periodicity means that there is no broadening. This situation is illustrated in Fig. 5.21(*c*). It is of course very possible that both types of disorder will occur together.

The decision that has to be made is whether to disqualify a layer structure from being smectic because it does not possess *long range* periodicity. De Vries [5.26] recognized this difficulty in the context of small-molecule liquid crystals and proposed a sub-classification called *cybotactic nematic* ($\kappa\acute{\nu}\beta o\varsigma$ = cube and $\tau\alpha\kappa\tau\acute{o}\varsigma$ = ordered). The term being reserved particularly for smectic C type structures in which the only diffraction maxima from the layers are first order, and diffuse. Such a pattern, recorded from a polymer of the series $\langle 60 \rangle$ with $m = 10$

$$\left[-O-\bigcirc\!\!\!\!-\underset{\underset{CH_3}{|}}{}N\!=\!N-\bigcirc\!\!\!\!-\underset{\underset{CH_3}{|}}{}O-\overset{\overset{O}{\|}}{C}-(CH_2)_m-\overset{\overset{O}{\|}}{C}- \right]_n \qquad \langle 60 \rangle$$

[5.27], oriented by drawing in the liquid crystalline phase and crystallized on cooling, is shown schematically in Fig. 5.22. The low

Fig. 5.21 Illustration of the effect which the reduction of order in smectic mesophases has on the diffraction patterns represented here as scans parallel to the z axis. The diffraction peaks from the layers are plotted as a function of the scattering vector, $\mathbf{s}_z$, which is related to the Bragg angle by: $|\mathbf{s}| = 2\sin\theta_B/\lambda$. The lines of vertical dots are linear lattices representing the relative positions of the smectic layers. (*a*) Smectic layers positioned on a one-dimensional lattice, and the associated diffraction pattern showing sharp periodic maxima. (*b*) Layers as before positioned on a lattice with a high degree of paracrystalline disorder; the diffraction maxima are much broader and only the first order is clearly delineated. (*c*) Layers positioned on a perfect lattice but within which the molecules are imperfectly arranged; the diffraction peaks are sharp but all high order maxima are suppressed.

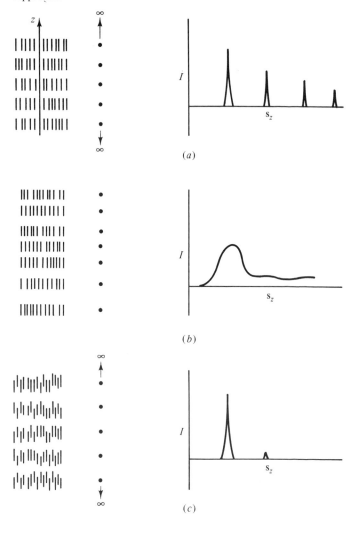

angle reflections are at an angle to the draw axis, indicating that the normals to the poorly formed layers are similarly oriented to the chain axis. The position of the maxima give a spacing normal to the layers of 1.58 nm at 50° to the chain axis. This spacing, when resolved in the chain direction, is 2.46 nm which compares well with chemical repeat of the fully extended molecule of 2.52 nm, and thus indicates that the flexible sequences are in a nearly extended conformation.

The term 'cybotactic' was first coined by Stewart and Morrow in 1927 [5.28], who used it to describe a level of positional order inferred from diffuse diffraction patterns of primary, normal alcohols. While the reasons for introducing a separate cybotactic nematic classification may be understandable, the fact that there is no qualitative difference between long range and short range order in one dimension, means that the association of the smectic classification with long range order is somewhat artificial, and there is no reason why smecticity should not include molecular segregation into layers which are less than perfectly spaced. This point of view has been clearly put by Atkins and Thomas [5.29], who suggest that the cybotactic nematic classification should be abandoned, at least in the context of layer structures in liquid crystals. Certainly, cybotactic nematics have led to considerable confusion in the field, and will not be referred to again in this text.

Fig. 5.22 Representation of the diffraction pattern from a smectic C polymer structure in which the layers are imperfectly ordered. (After [5.27] for polymer ⟨60⟩.) The director is vertical.

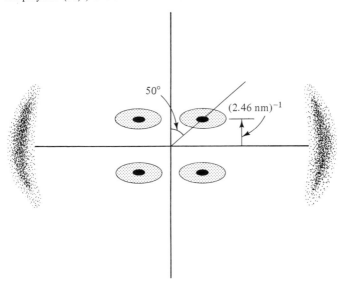

5.4.4   *Smectic stability in polymers*

(*a*) *Main-chain polymers*. The relative stability of nematic and smectic mesophases in polymers containing flexible sequences of units in the backbone, depends on the length of the sequence. And also, at least in the case of an alkane sequence, on whether it contains an odd or an even number of carbon atoms. In general, longer flexible sequences and odd numbers of carbon atoms favour the smectic phase over the nematic. Related trends are also apparent with sequences consisting of ethylene oxide units.

$$\left[-\overset{\overset{O}{\parallel}}{C}-\bigcirc\!\!-O-\overset{\overset{O}{\parallel}}{C}-\bigcirc\!\!-\overset{\overset{O}{\parallel}}{C}-O-\bigcirc\!\!-\overset{\overset{O}{\parallel}}{C}-O-(CH_2)_m\text{-}O- \right]_n \qquad <61>$$

For polymers of the type $\langle 61 \rangle$, Lenz [5.30] has shown that, when m is even and not greater than 6, the nematic phase is the only stable mesophase. At $m = 8$ and above the smectic phase only is stable, while with increasing n the temperature range of the mesophase decreases until it is only 33 °C at $m = 12$ and there is some suggestion that liquid crystallinity is completely absent at $m = 14$ or above. Odd values of m, in the range 3 to 9 reported, all show smectic phases over the full temperature range of liquid crystalline stability, with the transition temperatures somewhat lower than for the even series, as one would expect from odd–even effect (cf. Chapter 3). Blumstein *et al.* [5.31] have examined the series of polymers $\langle 60 \rangle$. It appears that the even members of the series show poorly ordered layers for molecules up to $m = 16$, while the smectic order is better developed for $m = 18$. The odd series shows an even greater reluctance to form layers although there is some evidence for them in sheared samples for which m is greater than 9. The situation regarding the effect of flexible sequences on smectic stability is far from clear. It *is* possible to say that increasing lengths of sequences favour smectic at the expense of nematic order, however it is difficult to make any similar generalizations regarding the odd–even effect.

(*b*) *Side-chain polymers*. Smectic polymers usually result from the segregation of the two parts of the side chain which are chemically distinct, the flexible spacers and the mesogenic rods. Intuitively, it may seem probable that smectics will be encouraged when the two types of sequence are present in roughly equal proportions, and the examples already quoted would tend to support this view. When the mesogenic rods are in the form of side chains, there are three locations of flexible sequence which contribute to smecticity. Figure 5.23 shows these in schematic form. The flexible tail (A) and spacer (B) contribute to smectic stability in much the same way as the tails of small-molecule smectogens as discussed above, however, the length of flexible backbone between side chains (C) is also important in the same sense.

5.4.5   *Optical textures in smectic polymers*

The high viscosity of polymer melts can make it difficult to achieve identifiable characteristic textures and this is especially the case with smectic polymers. Frequently a fine granulated texture of birefringent domains is seen between crossed polars such as that shown in Fig. 5.24(*a*) from polymer ⟨62⟩. This texture coarsens on annealing

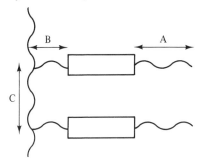

⟨62⟩

Fig. 5.23 Plot showing the effect of increasing the flexible chain content of a side-chain molecule, A + B + C, on the smectic stability. The stability is initially increased with respect to nematic, although the overall liquid crystalline stability is decreased.

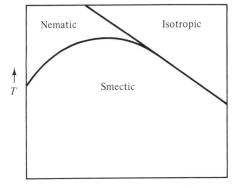

Increasing content of A + B + C ➞

into a *focal-conic fan texture* such as that illustrated for the same polymer in Fig. 5.24(*b*). The texture is characteristic of a smectic A phase, and its physical origins are described in Chapter 6. Another route to a focal conic texture is seen when a smectic forms directly on cooling an isotropic phase. The texture develops via the nucleation and

Fig. 5.24 Photomicrographs showing textures observed in smectic liquid crystalline polymers: (*a*) granulated texture; (*b*) focal conic fan texture. (Courtesy of Prof. V. Shibaev.)

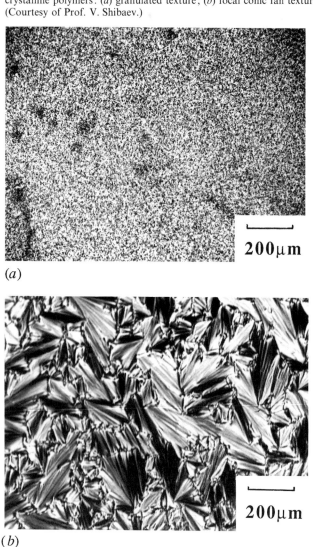

200μm

(*a*)

200μm

(*b*)

growth of characteristic bilateral focal conic entities known as *bâtonnets*.

The focal-conic fan textures are much less distinct in polymers in the smectic C phase where they are granulated. They are often called *broken focal conics*, and the micrographs of Shibaev and Platé [5.19] in

Fig. 5.25 The breaking up of the focal conic texture (*a*) of the smectic A phase of polymer ⟨47⟩ as it transforms to the smectic C phase (*b*). [5.19]. (Courtesy of Prof. V. Shibaev.)

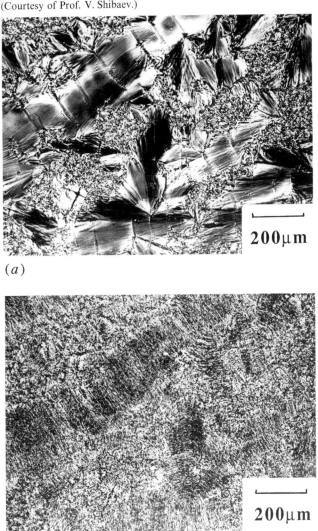

(*a*)

(*b*)

Fig. 5.25 clearly illustrate the textual transformation associated with the smectic A→smectic C transition at 30 °C in polymer ⟨47⟩ ($m = 11$).

5.4.6   *Chiral smectics in polymers*

The organization of chiral small-molecule liquid crystals into layer structures, where the molecular orientation precesses around the layer normal on translation parallel to this axis, has been mentioned in Chapter 2. It is sufficient to note here that similar C* structures have been observed [5.32] in mesogenic side-chain polymers such as ⟨63⟩ in which the flexible spacer contains more than six carbon atoms. The particular usefulness of chiral smectics is in the device field (see Section 8.5.1), and stems both from ferroelectric characteristics of the C* phase, and the ability to incorporate other types of active group into the structure by copolymerization.

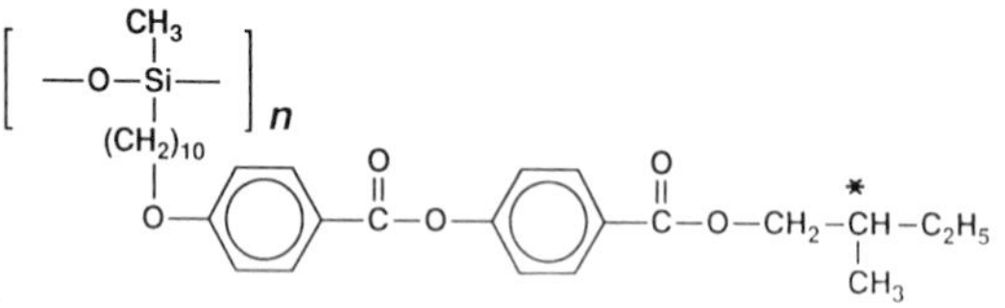

⟨63⟩

5.5   **Crystallization in liquid crystalline polymers**

5.5.1   *Effect of crystallization on orientation*

Liquid crystalline polymers for load-bearing structural applications are, of course, used in the solid state. The mesophase plays its essential role during processing, while solidification on cooling or solvent removal is often induced by the transformation to the crystal phase. Crystallization involves the local reorganization of the molecules into regions which have three-dimensional *positional* as well as orientational order. However, it is important to ask whether the local molecular orientation at a point within the crystal reflects accurately the orientation of the liquid crystalline polymer from which it formed. Or to put it another way, does the positional reorganization which occurs on crystallization change the local orientation of the director? It appears that although it is likely to do so in small-molecule materials, it does not in polymers. Figure 5.26 is an example of a random copolyester ⟨32⟩ of intermediate molecular weight with a degree of polymerization (DP) of 25, which has been oriented in the nematic phase in a magnetic field of 1.2 tesla along the vertical axis on the page, and crystallized on cooling in the field. The significance is that the diffraction peaks which dominate the diagram come from the crystalline component of the polymer, perhaps 20% of it. Yet their orientation faithfully reproduces the preferred alignment introduced into the melt by the field.

5.5.2 Crystals in main-chain liquid crystalline polymers

Transmission electron microscopy studies of Kevlar fibres [5.33], which crystallize as solvent is removed during fabrication, have shown significant evidence for crystallization. Not only can crystallites be observed in dark field, Fig. 5.27, but it has proved possible to image both equatorial and meridional lattice fringes at high resolution [5.34].

In the case of thermotropic random copolyesters, the issue of solid state structure is even more intriguing. At the outset, it is perhaps difficult to understand how copolymer chains with random sequences of units can contribute to the three-dimensional positional order of a

Fig. 5.26 Transmission X-ray diffraction pattern of a liquid crystalline copolyester $\langle 32 \rangle$ (DP = 25), cooled in a field of 1.2 tesla oriented vertically.

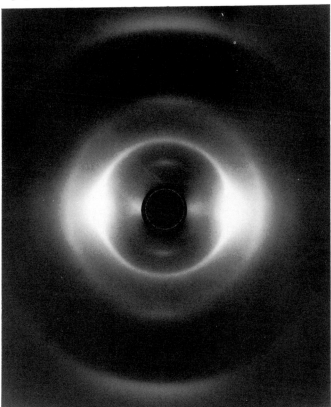

crystal, and yet there is ample evidence for crystallinity in many random copolyesters. Figure 5.28 collects some of the evidence for polymer ⟨32⟩. There is a DSC endotherm at the temperature (250 °C–320 °C, depending on composition) at which the polymer begins to flow as a melt, Fig. 5.28(*a*). There are sharp X-ray diffraction peaks corresponding to 10%–20% crystallinity which disappear on melting, Fig. 5.28(*b*). Crystallites can be imaged in dark field transmission electron microscopy where they appear as thin platelets, 10–20 nm thick, oriented normal to the local chain axes, Fig. 5.28(*c*). The platelet morphology is also confirmed by differential chemical etching of the crystallites, and subsequent imaging in the scanning electron microscope, Fig. 5.28(*d*).

X-ray diffraction patterns of fibre samples of this polymer show diffraction maxima spaced aperiodically along the meridian [5.37]. These are representative of the random distribution of the units along the chain, and yet their lateral concentration onto the meridional axis, especially in annealed samples, indicates that a particular random

Fig. 5.27 Crystallites in a Kevlar fibre imaged by dark field transmission electron microscopy. The mean chain axis is vertical, and the horizontal banding is associated with periodic changes in crystallite orientation about the vertical axis. The crystallites can be identified with the finer texture apparent in the chain direction. (Courtesy of Dr D. J. Johnson.)

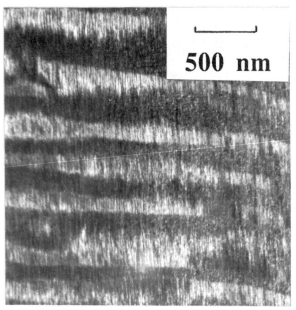

500 nm

Fig. 5.28 Evidence for the formation of crystallites in the random copolyester of hydroxybenzoic (73%) and hydroxynaphthoic acids (27%) ⟨32⟩. (*a*) DSC trace showing melting endotherm. [5.6] (*b*) Comparison of X-ray diffraction traces below (at 200 °C) and above (at 350 °C) the melting point. [5.35]

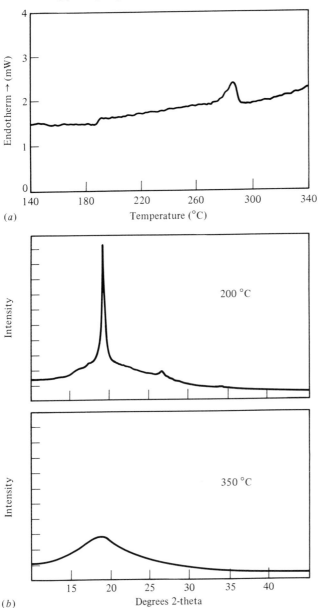

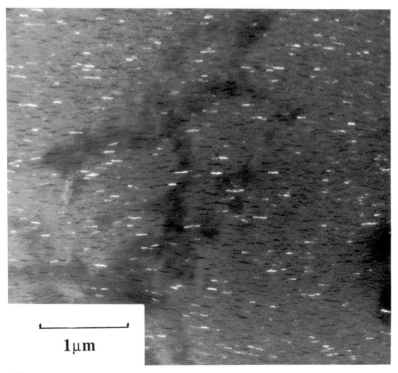

1μm

(*c*)

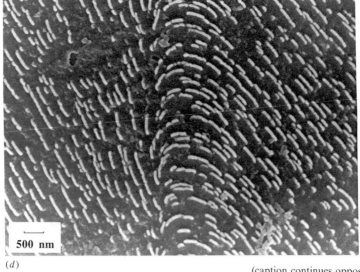

500 nm

(*d*)

(caption continues opposite)

sequence in one chain must be correlated with identical sequences in neighbouring chains as illustrated in Fig. 5.29(*a*). The crystals thus formed are in essence one (long) unit cell thick and are examples of 'aperiodic crystals' postulated by Schrödinger in 1944 [5.38] as a possible mechanism for the transmission of the genetic code. In the liquid crystalline polymer context they have been referred to as non-periodic layer (NPL) crystals. The idea of sequence matching [5.6] provides an explanation of the observed levels of crystallinity, the aperiodic but sharp meridional diffraction peaks, the crystal morphology, and, in terms of the possibility of diffusion of given sequences

Fig. 5.29. (*a*) A diagram showing the formation of a non-periodic layer crystal, by the juxtaposition of identical sequences along the random copolymer chains. (*b*) A computer model showing matching regions between chains which have been allowed to sort themselves by movement along the chain axes. The model consists of 100 chains each of 100 units of two different lengths. The chains are vertical in the figure and the regions of longitudinal register are clearly apparent. [5.39]

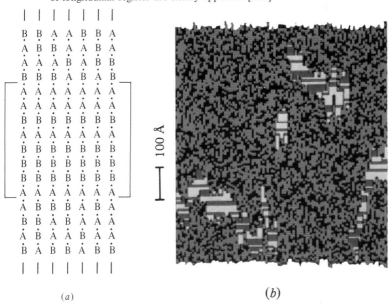

(*a*)   (*b*)

Fig. 5.28 (*cont.*)

(*c*) Crystallites imaged in dark field transmission electron microscopy with the chain axis vertical. [5.36] (*d*) Etched surface of a lower molecular weight version of the polymer, showing the crystallite platelets, and their reorientation across a domain boundary. (Courtesy of Dr T. J. Lemmon.)

towards the crystal growth front, the annealing effects. Figure 5.29(*b*) shows a computer model of NPL crystallites formed by the longitudinal segregation of random chains. Their size and morphology is similar to that observed microscopically.

6   Distortions and defects

6.1   Introduction

The *order parameter*, S, describes the quality of alignment of the molecular reference axes with the director. For a small-molecule liquid crystal in which the molecules have been aligned parallel to some external field, the order parameter will remain constant irrespective of the particular volume of material sampled, at least right down to molecular dimensions. However, in samples such as that showing the Schlieren texture in Fig. 5.4, the orientation of the director changes with position over distances of the order of microns. Viewed globally, there will be no preferred orientation and S will equal zero, at least in the viewing plane; but if only a small volume is sampled, S will be typical of the liquid crystalline phase, and, in the case of many polymer systems, quite close to unity. Thus, the measured value of the order parameter can depend on the volume sampled. In many circumstances it may be that the variations in director associated with the texture occur sufficiently gradually to provide a useful range over which sampling volumes are small enough to give a value which is characteristic of the molecular scale, but still sufficiently large to enable the measurement to be taken. An appreciation of the distinction between quality of orientation on a highly local scale, and the much lower degree of preferred orientation on a global scale – often as the result of many varied and beautiful textures which are a hallmark of liquid crystallinity – is crucial to any understanding of the liquid crystalline state.

Consideration of a diffraction pattern can provide insight into this distinction. Figure 6.1(*a*) shows an electron diffraction pattern of a thermotropic random copolyester $\langle 32 \rangle$ taken from an area $\sim 2.5\ \mu m$ in diameter. The positions of the equatorial maxima provide information on the nearest neighbour separations between chains. One might anticipate that, additionally, the length of the equatorial arc will determine the order parameter. Indeed it does yield a value, but Fig. 6.1(*a*) should now be compared with Fig. 6.1(*b*), a diffraction pattern of the same sample from an area now only $\sim 0.2\ \mu m$ in diameter (i.e., a volume smaller by a factor of about 150). This second diffraction pattern provides identical information on the nearest neighbour separation – the local packing – but the arc length of the equatorial reflection is clearly reduced, yielding a much higher value for *S*. The measured orientational order depends on the volume sampled.

In contrast to Chapter 5, this chapter will analyse the molecular organization on a scale much larger than the molecular level, and in particular it will examine the nature of defects. In fact most of the spatial variations in the director field which are considered here can be observed by optical microscopy and treated using continuum theories.

Fig. 6.1 (*a*) Selected area electron diffraction pattern of a random copolyester ⟨32⟩ taken from an area 2.5 μm in diameter. (*b*) Microdiffraction pattern of the same polymer recorded in the transmission electron microscope, taken from an area 0.2 μm in diameter. (After [6.1].) The director in each case is vertical.

(*a*)

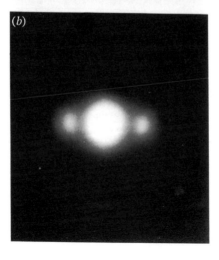

(*b*)

6.2 Elastic distortions

6.2.1 *Introduction*

When the director field varies continuously over a sample, a continuum theory is normally used to specify the distortions involved. For small-molecule liquid crystals, the typical length scale over which distortions occur is many times the molecular length and continuum models are entirely appropriate. For polymeric liquid crystals there is the additional factor that the elastic constants and distortion energy may be dependent on chain length. Indeed it is an experimental observation that the macroscopically measured parameter that plays the role of an elastic constant is dependent on the molecular weight and not just the chemical composition of the material, and this dependence can play a crucial part in determining resultant textures. For the moment, however, the simple continuum approach will be developed to serve as an introduction to the basic concepts.

6.2.2 *The three elastic constants*

For static distortions, three *elastic constants* can be identified. They are referred to as *splay* (K_1), *twist* (K_2), and *bend* (K_3). In some notations, they are identified by a double subscript as K_{11}, K_{22} and K_{33} respectively, and they are also referred to as the *Frank constants*. (It may assist memory of the assignment of K_1, K_2, and K_3 to Splay, Twist and Bend respectively to think of the word STaB.) The fundamental distortions involved in mesogenic chain systems are illustrated in Fig. 6.2(a–c), and are, in essence, very similar to those for small-molecule liquid crystals illustrated in Fig. 2.7. Also, as with small molecules, it should be kept in mind that any actual director field can be represented as an appropriate combination of these.

As outlined in Chapter 2, the distortions can be described in terms of deviations in the director **n**: for splay div **n** $\neq 0$; for twist, **curl n** is parallel to n; and for bend, **curl n** is perpendicular to **n**. The free energy density due to distortions can then be written in terms of the elastic constants as

$$F_{\mathrm{d}} = \tfrac{1}{2}(K_1(\mathrm{div}\,\mathbf{n})^2) + \tfrac{1}{2}(K_2(\mathbf{n}\cdot\mathbf{curl}\,\mathbf{n})^2) + \tfrac{1}{2}(K_3(\mathbf{n}\times\mathbf{curl}\,\mathbf{n})^2)) \quad (6.1)$$

All three elastic constants are necessarily positive, otherwise uniform orientation would not correspond to a state of minimum free energy. For small-molecule nematics it is generally found that twist has the smallest, and bend the largest value. All three are in the region of 10^{-12} N, and their values are temperature dependent. For small-molecule smectics, distortions which tend to alter the layer spacing are very high energy processes. Thus for smectics, twist and bend distortions are unlikely to occur, and this gives rise to their characteristic textures (focal conics) which are discussed further in Section 6.4.

For polymeric mesophases, the three elastic constants are functions

Fig. 6.2 The geometry of (*a*) splay, (*b*) twist and (*c*) bend distortions in a nematic liquid crystalline polymer.

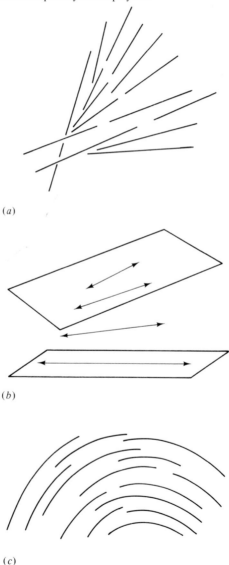

(*a*)

(*b*)

(*c*)

of chain length. De Gennes [6.2] was the first to point out that for long rigid rods, splay distortions are unlikely to occur. Figure 6.3 shows the necessity of segregating chain ends to achieve splay distortion while maintaining reasonable dense packing without overlap. Segregation may be hard to achieve for both entropic and kinetic reasons, particularly in fully dense thermotropics, and the problem will be exacerbated as the molecular weight tends to infinity. For lyotropics the situation may be somewhat easier, since the solvent molecules can assist in the maintenance of constant density in the vicinity of the distortion, and also increase the rate of diffusion of chain ends to the distorted region [6.3]. One might thus regard the elastic constants as a function of time, changing as diffusion permits the necessary segregation for splay to take place. Formally, it can be seen that without segregation of chain ends, splay must be associated with quite unacceptable density gradients in the material, since:

$$\operatorname{div} \mathbf{n} = -\mathbf{n}\frac{\Delta\rho}{\rho} \tag{6.2}$$

where ρ is the number density of chains cutting a plane perpendicular to $\mathbf{n}$.

It is difficult to envisage bend distortion with long, completely rigid rods on account of steric hindrance, although an attempt has been made in Fig. 6.4. However, real molecules, even those possessing no flexible spacers in the backbone, will often have sufficient flexibility to make bend possible. Thus even for fully *para* linked aromatics, the molecule can bend to follow the local curvature as long as the radius is greater than its persistence length (cf. Fig. 6.2(c)).

Twist distortions will not be impeded for the ideal case of a nematic possessing perfect order: successive planes of molecules will be stacked

Fig. 6.3 Configuration of chains involved in splay distortion, showing higher concentration of chain ends necessary to prevent overlap near to the 'origin' of the splay.

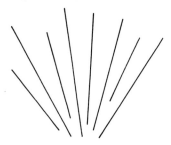

one on top of another (just as in a cholesteric). However, as soon as any disorder is introduced into the individual planes, molecules from one will protrude into a neighbouring plane possessing different alignment, causing a disturbance of the packing. As with bend distortions, this mismatched alignment can be accommodated by molecules of finite flexibility.

6.2.3   *Measurement of the elastic constants*

To date, little information has been gathered on the numerical values of elastic constants of liquid crystalline polymers. Methods developed for small-molecule mesophases cannot always be directly extended to polymeric systems [6.4]. The system that has been most extensively studied is the lyotropic polymer PBLG $\langle 22 \rangle$, and of the three elastic constants most attention has been given to K_2 (twist).

PBLG is a cholesteric with a pitch $\mathscr{P}$ typically of the order of a few tens of microns, although the precise value is both temperature and solvent dependent. As discussed in Chapter 5, when the helical axis lies in the specimen plane a 'fingerprint' texture is observed under the polarizing microscope (Fig. 5.12). The equispaced dark lines arise from planes where the molecules lie in the direction of observation, which occurs twice per pitch. Thus measurement of their spacing yields $\mathscr{P}/2$. If now a sufficiently strong magnetic (or electric) field is applied, the cholesteric helix may be 'unwound' to give an untwisted nematic. As the critical field strength, H_c, is approached the pitch increases, slowly at first, and then more rapidly. This effect is well known for small-molecule materials also, and the value of the critical field can be directly related to K_2 [6.4, 6.5] by the expression

$$H_c = \frac{\pi^2}{2\mathscr{P}} \left(\frac{K_2}{\chi_a} \right)^{\frac{1}{2}} \tag{6.3}$$

where χ_a is the diamagnetic anisotropy. (A similar equation holds for

Fig. 6.4 Ideally rigid rods are hindered from bend distortions by steric overlap. If the molecules are sufficiently flexible then this problem can be overcome, as shown in Fig. 6.2(c).

the critical electric field E_c with χ_a replaced by the dielectric anisotropy ε_a.) Figure 6.5 shows some experimental results obtained for PBLG in dioxane (20% wt/vol) [6.7].

It is found that K_2 is solvent dependent, but with a value not dissimilar to K_2 for small-molecule liquid crystals; for example, it has the value of 2×10^{-12} N for PBLG of molecular weight 310000 in chloroform. However, molecular weight now plays an important part so that an order of magnitude increase is found (albeit in a different solvent, methylene chloride) as the molecular weight is increased to 550000. Using the Onsager approach a relationship has been obtained between rod length L and K_2:

$$K_2 \sim 0.02 \; kT\rho^2 L^4 d \tag{6.4}$$

where ρ is the number density of rods and d the rod diameter, which predicts a very strong dependence on molecular weight. Insufficient data are currently available to confirm the accuracy of equation (6.4), but qualitative agreement is found.

For small-molecule liquid crystals it is possible to derive all three elastic constants independently by applying a magnetic field to a uniformly oriented nematic, although a different geometrical configuration is needed for each K in turn. Figure 6.6 depicts the three situations. In each case uniform orientation is required at the

Fig. 6.5 The dependence of the pitch $\mathscr{P}'$ of the cholesteric structure in PBLG (molecular weight 310000) in dioxane on the field strength H plotted in reduced coordinates; $\mathscr{P}$, the pitch in the absence of a field is 30 μm; H_c = 0.51 T is the critical field at which the pitch is infinite. The solid line represents the theoretical prediction [6.5, 6.6].

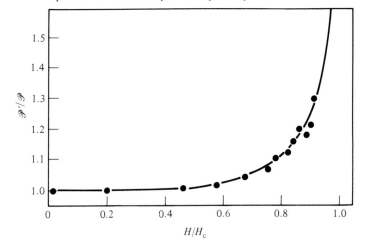

constraining surface: homogeneous for splay (Fig. 6.6(*a*)) and twist (Fig. 6.6(*b*)), homeotropic for bend (Fig. 6.6(*c*)). For small-molecule materials, this can usually be readily achieved by one of a variety of means such as rubbing, cleaning or deposition of an organic or polymeric film, but for liquid crystalline polymers the desired uniformity of orientation seems harder to achieve. For PBLG it seems that most preparations favour homeotropic alignment, although one route to homogeneous alignment has been reported in which a thin layer of poly(tetrafluoroethylene) (PTFE) is deposited on the glass surfaces.

The idea underpinning the method is attributed to Freedericksz [6.8, 6.9] (Frederiks or sometimes Fredericks in later papers), and is based on the competition between the alignment induced by strong surface anchoring and that induced by the external electric or magnetic field. For example, in the presence of a magnetic field an additional term

Fig. 6.6 The geometry for the three principal Fredericks deformations – (*a*) splay, (*b*) twist and (*c*) bend – for magnetic field directions indicated by the arrows to the left of the diagram. When fields are less than H_c surface anchoring dominates, but for $H > H_c$ distortion occurs as alignment along the field proceeds. In these diagrams the length of a rod indicates the length of a molecule projected onto the plane of the diagram.

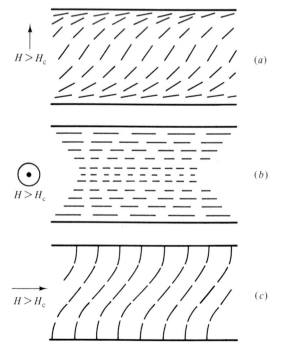

$\chi_a(\mathbf{n}\cdot\mathbf{H})^2$ appears in the expression for the free energy, and the total free energy can be reduced by a rotation of $\mathbf{n}$ towards $\mathbf{H}$. The consequent distortions shown in Fig. 6.6(a) (splay), 6.6(b) (twist) and 6.6(c) (bend) only occur above a critical field H_c given by:

$$H_c = \frac{\pi}{t}\left(\frac{K}{\chi_a}\right)^{\frac{1}{2}} \tag{6.5}$$

where t is the thickness of the cell. It has been found that the cholesteric helix of PBLG can be suppressed as a result of appropriate treatment of the glass surfaces to give homeotropic alignment combined with the selection of a cell whose thickness is less than the helical pitch in the bulk material. The result is a homeotropic nematic well suited for the measurement of K_3.

One of the major difficulties in applying this approach to liquid crystalline polymers arises from instabilities which cause periodic 'stripes' associated with systematic deviations in director orientation. However, it has been shown, [6.4] and [6.10], that these instabilities can be turned to good advantage as they can provide an alternative means of measuring the elastic constants, as well as the viscosity coefficients. This approach is discussed as part of a fuller treatment of field effects in Chapter 7.

Although comparatively few measurements of elastic constants have been made on polymer systems, data are available for PBLG [6.4, 6.11 and 6.12] and for a main-chain thermotropic polyester with a pentamethylene spacer [6.10]. All reports agree on K_2 (twist) being the smallest of the three elastic constants, but it appears that the relative magnitudes of splay and bend depend on the particular system. For example, [6.11] and [6.12] report that the splay elastic constant is much greater than bend, by up to three orders of magnitude in the case of PBLG in dioxane with 2% trifluoracetic acid. In contrast a racemic mixture of PBLG and its enantiomer PBDG in a dioxane/methylene chloride mixture gives a K_1 that is marginally smaller than K_3 [6.4]. It is not clear whether this difference can simply be attributed to the choice of solvent, and clearly more data are needed to resolve the important question of the relative magnitudes of the K's. The only data on a thermotropic material, in agreement with [6.11] and [6.12], indicate that K_1 (splay) is much larger than K_3 (bend), in fact by an order of magnitude. K_3 in turn is only slightly greater than K_2 (twist) [6.10].

6.2.4   *Splay compensation*

Because the splay elastic constant is the largest of the three Frank constants, and especially large in polymer systems, the distortions that occur will tend to avoid splay. However, Meyer [6.13] has pointed out that the problem of macroscopic density variations associated with splay distortion of long rod molecules, can be alleviated by splay of an opposite sign in an orthogonal plane. The essence of *splay compensation* is illustrated in Fig. 6.7. The splay in the x–z plane is $(\partial n_x / \partial x)$, while that in the y–z plane is $-(\partial n_y / \partial y)$. These opposing distortions will tend to compensate, and if they are of equal magnitude the total splay distortion will be zero, and the density variation along the z axis will be eliminated without the need for extensive segregation of chain ends. The organization of such regions to fill space must necessarily lead to discontinuities in the structure. Examples of splay compensation have been observed in both lyotropic (PBLG) and thermotropic systems ⟨32⟩.

6.3   **Disclinations**
6.3.1   *Geometry*

So far in this chapter the director has been assumed to vary smoothly and continuously throughout the sample. However, in practice *singularities* in the director field may occur. These are known as *disclinations*, and may be point or line defects. Whereas in a crystal a dis*loc*ation can be regarded as a discontinuity in translation, in a liquid crystal a disclination is a discontinuity in orientation.

Disclinations are responsible for many of the textures seen under the polarizing microscope, such as the *Schlieren* texture, otherwise known as *structures à noyaux*, which is shown in Fig. 5.4. A characteristic appearance of a disclination between crossed polars is a stationary

Fig. 6.7 Illustration of splay compensation. There are two components of splay represented on orthogonal faces, which because of their opposite sign will tend to compensate. In the case of equal and opposite values, the high energies associated with density variations along z will be eliminated.

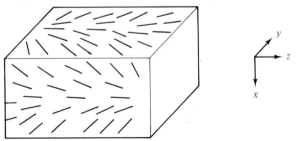

Fig. 6.8 (*a*) Identification of the angles θ and Φ (in polar coordinates) for describing a disclination. (*b*)–(*g*) Schematic representations of the molecular arrangement at disclinations with different values of *s* and *c*.

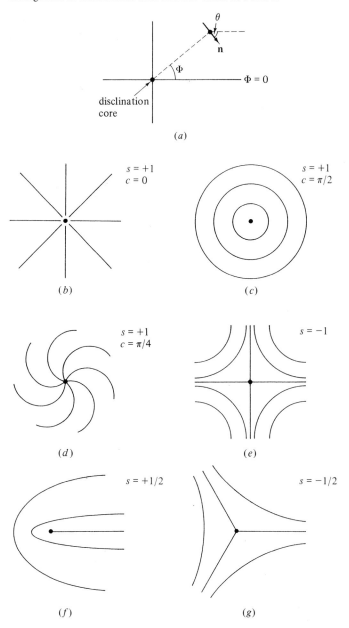

(*a*)

$s = +1$
$c = 0$

(*b*)

$s = +1$
$c = \pi/2$

(*c*)

$s = +1$
$c = \pi/4$

(*d*)

$s = -1$

(*e*)

$s = +1/2$

(*f*)

$s = -1/2$

(*g*)

point singularity from which two or four dark 'brushes' (extinction lines) radiate. The brushes rotate as the polars are rotated. The distortion field around a disclination can be defined by two parameters θ and Φ. These parameters are shown in Fig. 6.8(*a*), where Φ represents the angular coordinate of a given point in polar coordinates with respect to the disclination core, and θ the angle the local director axis at that point makes with the $\Phi = 0$ axis. Consideration of the free energy of distortion for the situation when all three elastic constants are equal e.g. [6.14], gives a relation for θ:

$$\theta = s\Phi + c \qquad (6.6)$$

where c is a constant and s is known as the *strength* of the disclination which may be positive or negative. It has already been shown that, while the 'one-constant' approximation is not particularly good for small-molecule liquid crystals, it is particularly poor in the polymeric case where the splay elastic constant (K_1) is large compared with twist and bend. With or without this approximation, there will be four positions of extinction in the polarising microscope per 2π rotation of the director, so equation (6.6) leads to a relation between the disclination strength s and the number of brushes seen radiating from it:

$$s = \tfrac{1}{4} \times (\text{number of brushes}) \qquad (6.7)$$

Figures 6.8(*b–g*) show the director fields at disclinations of strengths of $s = \pm 1$, $\pm\tfrac{1}{2}$. These director maps may be regarded as relating either to point defects, or line defects lying normal to the page.

If a complete circuit is made around the centre of the $|s| = \tfrac{1}{2}$ disclination, the director rotates by π. The sense of the director rotation is, however, opposite for the $s = +\tfrac{1}{2}$ and $s = -\tfrac{1}{2}$ cases. Specifically, the rotation of the director is in the same sense as the sense in which the circuit is traversed for an $s = +\tfrac{1}{2}$ disclination, but in the counter sense for $s = -\tfrac{1}{2}$. For $|s| = 1$, a similar circuit yields a total director rotation of 2π. In this case also the sense of rotation follows that of the circuit taken for the positive strength disclination, and is opposite to it for the negative case. This pattern of behaviour can be verified by examination of the director maps drawn in Fig. 6.8(*b–g*).

Equation (6.6) clearly allows the possibility of s exceeding 1 and c taking a range of values $0 < c < 2\pi$. The five configurations shown in Fig. 6.8 seem to be the only important cases in practice and disclinations of strengths greater than 1 have never been observed; the energy of a disclination is proportional to s^2, and the associated energy would simply be too great. Likewise the only important non-trivial values for c seem to be $c = \pi/4$ and $\pi/2$ for the $s = +1$ case (Fig. 6.8(*c*) and (*d*)).

Using these director maps the appearance of an isolated disclination under the polarizing microscope can now be understood. The dark 'brushes' emanating from a singularity (Fig. 6.9(a)) correspond to lines of extinction, where the local extinction axis lies parallel either to the polarizer or (crossed) analyser. As the crossed polars are rotated (Fig. 6.9(b)), the brushes rotate. For |s| = 1 disclinations, extinction

Fig. 6.9 (a) Appearance of disclinations in a random copolyester ⟨33⟩ between crossed polars. (b) Effect of an anticlockwise rotation of 55° of the crossed polars on the contrast. (c) Director map constructed from (a) and (b) and related photographs. Note that there is a 90° uncertainty in the molecular orientation at any point, and a second map in which the streamlines are at all points normal to those here would be equivalently valid on the basis of the evidence available. [6.15]

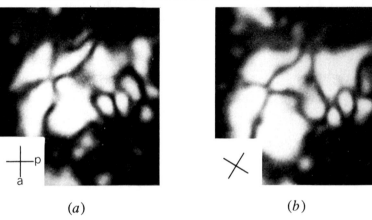

(a)   (b)

(c)

occurs along four radial lines, for $|s| = \frac{1}{2}$ along two. As the crossed polars are rotated the brushes of an $|s| = 1$ disclination will rotate by the same amount, either in the same sense (*s* positive), or in the opposite (*s* negative); the two brushes of an $|s| = \frac{1}{2}$ disclination rotate twice as fast as the crossed polars. From the study of the behaviour of the brushes as the polars are rotated, the director field can be mapped out, and the type of disclination therefore identified. However, because the use of crossed polarizers cannot distinguish the case of the extinction axis lying parallel to the analyser from that of this axis being parallel to the polarizer, a 90° degeneracy remains, and two orthogonal maps can always be drawn in the absence of any supplementary information such as might be provided by boundary conditions (Fig. 6.9(*c*)).

The picture of disclinations described here has been built up over the years in relation to small-molecule liquid crystals. However, two caveats are in order, particularly where an extension to polymeric materials is being considered. First, as has been discussed in Chapter 5, there is evidence that a number of liquid crystalline polymers may be optically biaxial. One consequence of this type of behaviour is that the direct identity between the molecular chain axis and the extinction directions for polarized light may be lost. For these materials there may be decoupling between the molecular chain axis and the extinction directors. Thus the director maps of Fig. 6.8 should not be assumed to correspond necessarily to molecular orientations, and care must always be taken in this respect when interpreting polarized light micrographs of mesophase textures.

The second problem has already been touched upon: the formulation has been set up within the one-constant approximation. When the three elastic constants are markedly different, deviations in the form of the distortions around the disclination core occur, and the brushes no longer rotate uniformly upon rotation of the polars. In principle this should provide information on the elastic anisotropy, but in practice it is difficult to achieve.

6.3.2   *Point, line and surface disclinations*

In discussing Fig. 6.8 it was stated that the geometrical representations corresponded either to point disclinations, or line disclinations viewed end on. Often line disclinations do lie perpendicular to the plane of the specimen because of interactions with constraining surfaces. When this occurs they can usually be revealed to be lines by slightly displacing the coverslip of the microscope specimen cell so that the intersections of the disclination line with the two surfaces no longer exactly superimpose. The threaded *structure à fils*, from which nematics acquired their name, corresponds to a texture full of disclination lines inclined to the surfaces. Their dark appearance arises from the discontinuity in optical properties along each line, producing phase contrast.

For small-molecule liquid crystals, it is now believed that line disclinations of strength $s = 1$ cannot in general exist because of 'escape in the third dimension' which is illustrated in Fig. 6.10 [6.16, 6.17]. The singularity at the core of the disclination can be relieved by a suitable reorientation of a planar structure into a structure in which the molecules tilt further away from the specimen surface the deeper one moves into the specimen, until they lie along the direction which would have corresponded to the core-line of the disclination. Thus the line singularity is relieved at the expense of a splay–bend distortion, to leave a point singularity; this can always be achieved if the energy associated with splay–bend is not too high. In practice, this situation holds for most small-molecule nematics.

Observations of a main-chain thermotropic nematic polymer have shown that $|s| = \frac{1}{2}$ disclinations greatly predominate over the $|s| = 1$

Fig. 6.10 (a) $s = +1$ disclination line lying along the z axis, plan view. (b) Side view of disclination showing surface of sample. (c) Same view as (b) but after 'escape' to give a point singularity on the surface. The planar view (a) will be unchanged.

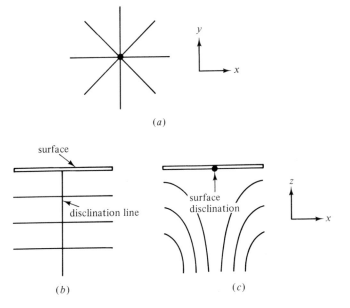

(a)

surface

disclination line

(b)

surface
disclination

(c)

type. This situation would be expected on the basis that the disclination energy is proportional to s^2, but not if the $|s| = 1$ disclinations are able to reduce their energy through escape. The apparent absence of escape can be explained in terms of splay distortion, an integral part of the process, having an especially high energy for the polymer.

So far it has been assumed that the disclinations reside in the bulk of the material, although a disclination line may be pinned at the surface. It is, however, possible for a *surface disclination* line to form in preference to point disclinations located at a surface. An understanding of surface disclinations owes much to the work of Klèman and coworkers, which is described in [6.18]. Unlike a disclination in the bulk, a surface disclination can have any strength s. This fact can easily be seen by performing the following 'thought experiment'. Imagine a solid containing two crystals, of arbitrary orientation, separated by a grain boundary (Fig. 6.11(a)), and let the material melt to a nematic mesophase. The free energy will be minimized by a relaxation of the orientation of the molecules within the bulk, but a line discontinuity

Fig. 6.11 The generation of a surface disclination. An arbitrary misorientation $\Delta\theta$ between two domains (a) becomes smoothed out (b) into a surface disclination line of strength $s = \Delta\theta/2\pi$.

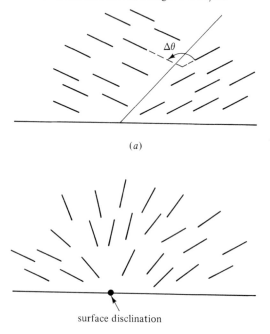

(a)

surface disclination

(b)

will remain at the surface, as it is assumed that there will be a considerable energy cost associated with the molecules meeting the surface at anything other than the preferred angle. The surface disclination corresponds to the intersection of the original boundary with the surface (Fig. 6.11(*b*)). Since the original misorientation of the two crystals was arbitrary ($\Delta\theta$), this surface disclination can be of arbitrary strength, related to $\Delta\theta$ by $s = \Delta\theta/2\pi$.

The strength of influence of the surface in inducing preferred orientation is described in terms of an *anchoring energy*, W_s, which is the energy required for the molecule to lie along the most unfavoured direction at the surface. For example, for the case of uniform orientation, the energy W associated with a molecule inclined at an angle θ to the surface is given by [6.18]

$$W = W_s \sin^2 \theta \tag{6.8}$$

There are thus two opposing influences. The distortion in the wall will involve an energy density K/t^2 acting over the distance t. For a sample thickness of h and a unit length of wall in the plane of the surface, the total distortion energy will be Kh/t. Hence the greater the thickness of the wall, the lower the energy. On the other hand, the existence of a wall of thickness t at the surface will cost an energy of tW_s. Minimizing the total energy with respect to t gives a thickness for the wall

$$t = \left(\frac{hK}{W_s}\right)^{\frac{1}{2}} \tag{6.9}$$

and a total energy E_w

$$E_w = 2(hKW_s)^{\frac{1}{2}} \tag{6.10a}$$

or

$$E_w = 2K\left(\frac{2h}{b}\right)^{\frac{1}{2}} \tag{6.10b}$$

where the *extrapolation length* b is defined by

$$b = K/2W_s \tag{6.11}$$

Since the energy associated with two surface disclinations together with a smeared out (diffuse) wall is $\sim 2K/\text{unit length}$, it can be seen that walls will only be favoured for $b > h$. For small-molecule liquid crystals this is hard to achieve because $W_s \sim 10^{-3}$ J m^{-2} leading to the need for film thicknesses of less than 10 nm. However for liquid crystal polymers, although measurements of W_s have not been made, it seems that walls may be somewhat easier to achieve. Klèman *et al.* [6.19] and Donald *et al.* [6.20] have observed them in two very different thermotropic nematics of thickness ~ 5–10 μm by optical microscopy. Furthermore, molecular weight effects are also likely to be important because of the difficulty of accommodating splay as the chains become longer. They have been noted in transmission electron microscopy studies of 100 nm thick films of a thermotropic copolyester ⟨32⟩, in which the walls develop upon annealing on a rocksalt substrate which promotes homeotropy [6.21].

6.3.3   *Observations of disclinations in nematics*

There have been frequent reports of disclinations, revealed by optical microscopy, in polymeric liquid crystals, both lyotropic and thermotropic, nematic, smectic and cholesteric. However, differences have been reported as to the details of the disclination structure and the conditions under which they form, and several outstanding questions remain as to the nature of disclinations and in particular the detailed core structure. As has been discussed in the previous section, escape in the third dimension to relieve the core singularity does not seem to occur in thermotropic polymers.

Millaud *et al.* [6.3, 6.22] observed integral disclination lines in a series of lyotropic solutions of poly(*p*-phenyleneterephthalamide) ⟨44⟩. The disclinations formed over a time, which increased with molecular weight, and were observed to be very fine. The explanation provided by the authors was that the disclinations required chain end segregation at the core, and the necessary diffusion took time to occur; the diffusion would be easiest to achieve for systems containing many chain ends, i.e. short chains. The assumption was that bend distortions were the highest energy distortions and therefore tended to disappear with time.

In thermotropic systems, chain end segregation is likely to be much harder to achieve, and the necessary diffusion times consequently prohibitively large. Thus rather thick disclination lines may be seen and, as observed by Klèman *et al.* [6.19], half integral strength lines are favoured. Klèman *et al.* have more recently extended their studies to look in detail at the disclination core [6.23]. They differentiate between $s = \frac{1}{2}$ and $s = -\frac{1}{2}$ disclinations. For the former they identify a configuration which does contain a large number of chain ends at the core, and with essentially no splay deformation away from the core. However, the $s = -\frac{1}{2}$ disclination can minimize elastic energy by adopting a structure with a large bend component but little chain end segregation. The differences in core structure lead to differences in mobility between the two types of disclination, the positive sign disclination being much less mobile because of the requirement that all chain ends must be dragged along if the singularity is to move. Transmission electron microscopy of $s = \frac{1}{2}$ disclinations in two other thermotropic polyesters [6.24] has been able to examine the anisotropy of the elastic constants – values for the parameter $(K_1 - K_3)/(K_1 + K_3)$ of -0.13 and 0.20 were obtained when the only change in chemistry was the replacement of a hydrogen atom by a methyl group on the hydroquinone moiety in the main chain. It is therefore clear that comparatively minor changes in chemistry can have significant effects on the elastic constants and hence the nature of defects in these polymers.

Surface disclinations and walls have also been observed in thermotropic polymers. Walls accommodate a difference in orientation between one domain and its neighbour and, although their thickness may be as little as a few tens of nanometres, the reorientation within them is continuous. Klèman *et al.* [6.19] reported both in a sample of a rigid–flexible spacer thermotropic polymer. The surface disclinations formed under conditions of strong planar anchoring, and were tied to a bulk disclination line. The shape of the resultant disclination loops, formed during capillary flow of the polymer into the cell, permitted an estimate to be made of the K_1/K_3 ratio at around 3.

In these same samples, walls formed during a transformation from a

Fig. 6.12 Schematic representation of the geometry at a wall with a 90° rotation of the molecules across the wall. (After [6.18].) The nail convention is used.

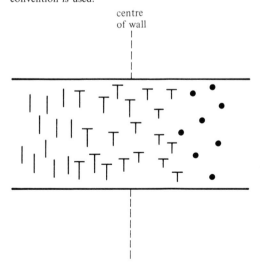

planar to a homeotropic texture. The walls separated transformed and untransformed regions, with a consequent 90° rotation of the molecules across them. The proposed geometry, akin to a 90° Bloch wall in ferromagnets in which the axis of rotation is normal to the plane of the wall, is shown in Fig. 6.12. Bloch walls are discussed by Klèman in [6.18]. Walls have also been observed in thin films of a main-chain thermotropic polymer, as revealed by transmission electron microscopy [6.21]. In this system the walls also developed as the molecules attempted to transform to a homeotropic alignment (on a rocksalt substrate), but the geometry was more akin to the 180° Bloch wall of ferromagnets. In the development of the homeotropic texture, molecular weight is clearly an important parameter since chain ends must condense at the surfaces, and this effect is apparent in the development of walls: only if sufficient low molecular weight material is present can homeotropy occur. Thus it seems likely that low molecular weight material is actually segregating in the wall. In this system the walls do not separate homeotropic and planar regions, but rather regions where the molecules in both domains have some finite misorientation with respect to the plane of the sample, but in opposite senses. In this case, the domain structure adopted seems to be an example of splay compensation in operation, with the splay distortions in one plane being compensated by a negative splay component in an orthogonal plane.

The wall thickness in these specimens is only a few tens of nanometres thick, and hence unresolvable by optical microscopy, but rather similar observations have been made on relatively low molecular weight samples of another main-chain thermotropic copolyester ⟨64⟩

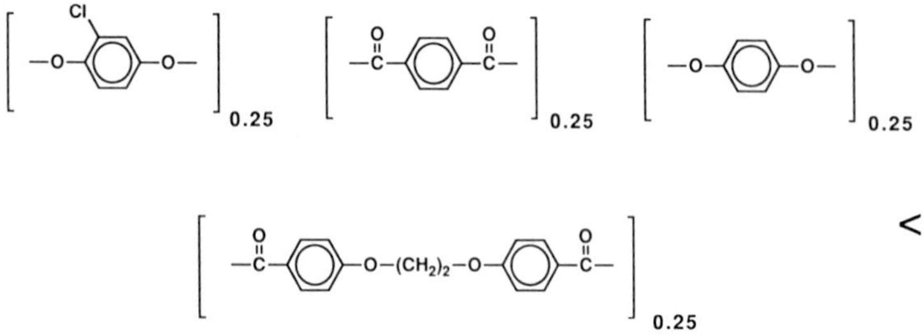

⟨64⟩

[6.20]. Once again, molecular weight is clearly a crucial parameter in determining the textures observed, since higher molecular weight samples of the same chemistry do not appear to show the same wall structures.

6.3.4 Observation of disclinations in smectics and cholesterics

The characteristic textures of the layer structures of cholesterics and smectics are determined more by the requirements of continuity of essentially undeformed layers than by the presence of individual disclinations. The resultant textures, often termed *focal conics*, will be discussed in Section 6.4.

There is an extensive literature on disclination lines in small molecule cholesterics. Because of the twisted nature of the phase, the lines are more complex than in a simple nematic and there are several variants depending on the axis of rotation. For further details the interested reader should turn to [6.18, 6.25]. There have been few comparable studies on liquid crystal polymers, but they have been shown to occur in solutions of DNA and xanthan (a natural polysaccharide).

Dis*loc*ations are thought to give rise to one of the characteristic textures of cholesterics, the so-called *oily streak* texture, in which the oily streaks are composed of rows of edge dis*loc*ations. Figure 6.13 shows an example of the oil streak texture in a chiral thermotropic polyester. It has been used as a means of confirming the existence of a cholesteric structure.

Dis*loc*ations and disclinations are also found in smectics. In polymers the clearest evidence for disclination structures has been produced by electron microscopy using a 'decoration' technique [6.27] (Fig. 6.14).

Fig. 6.13 Cholesteric texture with oily streaks. (After [6.26].) (Courtesy of Prof. E. Chiellini.)

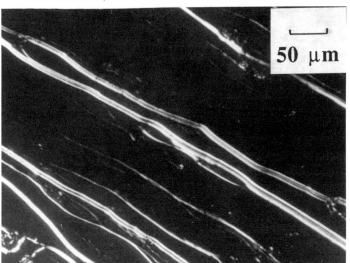

50 μm

In this method the thermotropic mesophase structure, a main-chain polymer with a decamethylene flexible spacer, is allowed to crystallize partially. The crystalline regions, which diffract strongly and therefore appear dark, run normal to the local chain axis. The micrographs, with their 'map' of director orientations, therefore provide a direct visualization of any discontinuities in the director field. Both $s = \frac{1}{2}$ and $s = 1$ lines have been observed in this way, and also a disclination dipole. The appearance of dis*loc*ations in a smectic phase has also been shown using transmission electron microscopy. Using a high resolution

Fig. 6.14 Disclinations in a smectic liquid crystalline polymer revealed by 'decoration' following partial crystallization. (Courtesy of Dr B. A. Wood.)

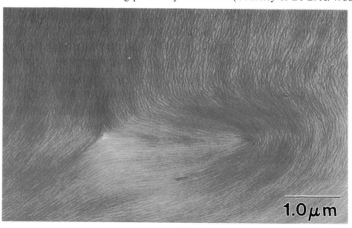

1.0μm

Fig. 6.15 The schematic appearance of smectic planes in the transmission electron microscope. The planes are seen to undulate, and dis*loc*ations can be identified where a smectic plane terminates. (After [6.28].)

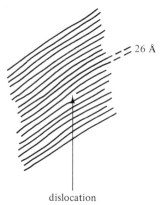

26 Å

dislocation

technique, discontinuities in the positional ordering of smectic planes, as sketched in Fig. 6.15, have been seen in thermotropic polymers. This technique has also been able to demonstrate that the smectic planes are undulating, with a characteristic wavelength of 0.1 μm.

6.4 Focal conics

Consider a smectic A mesophase based on a main-chain liquid crystalline polymer with flexible sequences in the backbone. Figure 6.16 shows the consequences of distortions in such a structure. Bend distortion of the layers must necessarily be associated with splay of the chain molecules, which is particularly unfavourable in polymers. Correspondingly, bending of the molecules, in itself not a particularly high energy process for polymers, will introduce splay into the layers which will be opposed by their resistance to dilation or compression. Additionally, twist distortion of the molecules will introduce substantial shears into the layers remote from the twist axis. The consequences of these rather severe restrictions on distortion have been observed in many smectic and cholesteric materials, where they lead to a family of characteristic textures called *focal conics*.

An understanding of the basic geometry which controls these textures has come from the small-molecule literature; only a brief description will be given here. More detailed discussions are given by

Fig. 6.16 Distortions associated with the layers. (*a*) Bend distortion of the layers will give splay of the polymer molecules, while (*b*) bend distortion of the molecules will give splay of the layers.

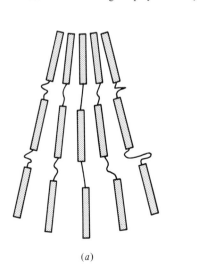
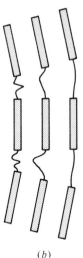

(*a*) (*b*)

Fig. 6.17 Dupin cyclide construction. (*a*) Vertical, and, (*b*) horizontal cross sections of cyclide; (*c*) vertical section showing layers of the structure; (*d*) horizontal section. Heavy lines in (*c*) and (*d*) indicate the ellipse, hyperbola, Dupin cyclide, and central domain. (After [6.29].) (*e*) Diagram of central domain showing structural layers, with a representation of the arrangement of molecules within one of them. (Redrawn from [6.30].)

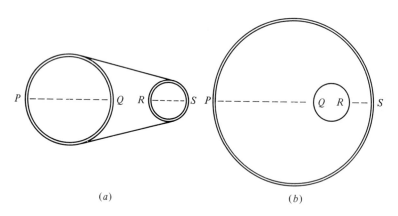

(*a*)   (*b*)

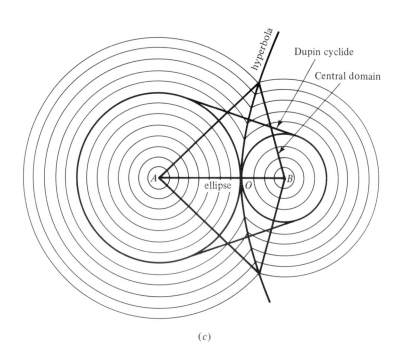

(*c*)

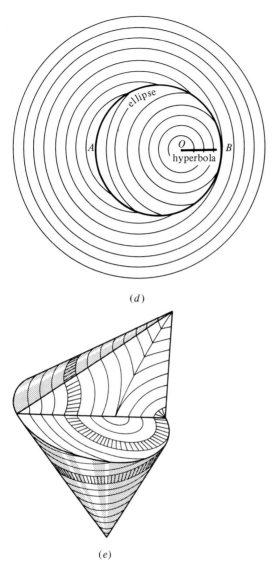

(*d*)

(*e*)

Hartshorne and Stuart [6.29], Klèman [6.18], and Gray and Goodby [6.30] who also present a significant series of micrographs. Examples of focal conic textures in polymer systems are reproduced in Figs. 5.24 and 5.25.

Analysis of the optical textures shows that the layers are deformed in a complicated way to give what is known as a *Dupin cyclide*. The core of the cyclide is shown in Fig. 6.17(*a, b*). The ring of the cyclide has everywhere a circular cross section, whose diameter may vary continuously. As successive smectic layers are added to the cyclides (Fig. 6.17(c, d)), those centred around A and those round B first touch at O, and a locus of these points of intersection may be drawn. In the principal section of the cyclides shown in Fig. 6.17(c), these points lie on a hyperbola, whereas in the orthogonal plane (Fig. 6.17(d)), the discontinuities lie on an ellipse. The vertex O of the hyperbola lies on the line joining A to B, and is one of the foci of the ellipse. The focus of the hyperbola is B, and A and B are the vertices of the ellipse. Thus the two curves are known as focal conics. Since the hyperbola and the ellipse represent discontinuities in structure, they can be directly seen under the polarizing microscope. If the cyclide has the same cross-sectional diameter all the way round, instead of varying continuously, then the ellipse becomes a circle and the hyperbola a straight line. It is important to appreciate that *complete* Dupin cyclides are never observed in mesophase textures and the structure is built up from a series of central domains such as that outlined in Fig. 6.17(c), and drawn in Fig. 6.17(e). It is significant that within this domain, the layers have a double opposite curvature. This shape will enable the splay associated with one curvature to compensate that associated with the other curvature.

Focal conics can lead to a variety of textures, as seen in the previous chapter. In the polygonal texture (described originally by Friedel [6.31]) half of each hyperbola is missing (either above or below the plane of the ellipse) and the domains are conical. The ellipses themselves are situated on one or other of the surfaces, their apices on the other. Thus space-filling can be accomplished by the juxtaposition of the central domains of Fig. 6.17(c), and this is represented schematically in Fig. 6.18(a). The cones themselves are packed so that the ellipses both touch each other tangentially and also are tangential to polygons which enclose them and this can be seen in Fig. 6.18(b). Clearly there are two families of polygons involved, one lying in each surface. The two families are orthogonal to one another (Fig. 6.18(c)), and each can be observed in turn by focusing first on one surface and then on the other. When cooling from the isotropic melt it is usually found that bâtonnets – anisotropic rod-like domains – form, which upon further cooling grow into fan-shaped textures (*plages en éventail* in Friedel's original terminology). The dark lines visible as the 'pleats' of the fan appear to correspond to hyperbolae viewed end on.

The above understanding has been derived from intensive studies of small-molecule materials, and there is a wealth of textures which qualify for the name 'focal conic'. These are copiously illustrated in the book by Gray and Goodby [6.30], and recent studies have shown that there are close similarities in polymer systems. For instance, Krigbaum and Watanabe [6.32] have observed the structural change isotropic liquid → bâtonnets → fan-shaped texture on cooling a thermotropic

Fig. 6.18 (*a*) The central domains shown in Fig. 6.17(*c*) and (*e*) can be packed together to fill space. (*b*) The packing of ellipses gives rise to polygonal structures (such as *EFGH*), the boundaries of which are tangential to the ellipses. *A*, *B*, *C* and *D* are the foci of the ellipses. (*c*) The polygons at the upper and lower surfaces are orthogonal so that on the opposite surface from (*b*), the polygon *EFGH* is out of focus as well as being displaced. Four in-focus polygons meet at *X*. (After [6.30].)

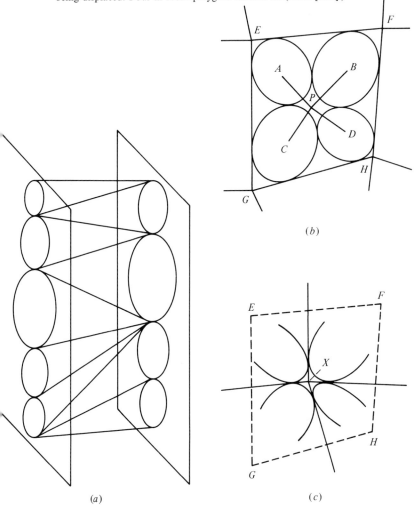

main-chain homopolyester with hexamethylene spacers. Noel *et al.* [6.33] have identified different smectic phases (S_A, S_C and S_E) from detailed observation of the appearance of fans, while fan-shaped textures have also been observed in polymers with mesogenic side groups [6.34, 6.35].

In all these focal conic textures it is clear that the surfaces play an important role in determining the texture. Although surface treatments are of paramount importance in the successful exploitation of liquid crystals in display devices, much less is known about surface interactions of mesogenic polymers (although see Section 6.2 above and Chapter 7) and the textures seen, useful as they are as an aid to phase identification, are not predetermined. Little has been done systematically to characterize the formation of the different textures in smectic and cholesteric liquid crystalline polymers, and it is not as yet clear whether the conventional liquid crystal literature is an adequate guide to the problem, although it clearly provides the necessary first steps.

6.5   Biphasic structures

As explained in Chapter 4, biphasic structures are to be expected in lyotropic solutions over a range of concentrations when the anisotropic phase occurs in conjunction with the isotropic. Figure 6.19 shows a typical biphasic structure in which 'spherulites' of anisotropic material are suspended in an isotropic matrix which appears black under crossed polars. This general appearance seems common to a wide range of lyotropics, including poly(γ-benzyl-L-glutamate) ⟨22⟩, poly(*p*-phenylene-terephthalamide) ⟨44⟩ and numerous cellulose derivatives such as ⟨20⟩ and ⟨21⟩.

The structure within the spherulite has been analysed. The conclusion is that the spherulite is 'negative' in optical terms, the tangential refractive index, n_θ, being greater than the radial refractive index, n_r. Since the chain polarizability is greater along the chain length rather than perpendicular to it, the conclusion is that the structure comprises macromolecules oriented tangentially. If the liquid crystalline polymer is also cholesteric, the fingerprint texture may be superimposed. The total effect is therefore of a sphere, showing a Maltese cross of extinction with the possible superposition of dark rings arising from the fingerprint texture.

For many thermotropic systems, particularly the rigid main-chain type, the liquid crystal–isotropic transition temperature is so high that polymer degradation sets in before the isotropic phase forms so that coexisting anisotropic and isotropic phases are not seen. However, this need not be so, in which case a biphasic texture may develop. Stupp and colleagues have studied the relationship between chemistry and the occurrence of biphasic structures in some detail, with particular emphasis on the role of chemical heterogeneity [6.36, 6.37]. They discovered that when the chain was chemically regular, the temperature range over which both isotropic and liquid crystalline phases coexisted was ~ 5°, and the texture over this range was very fine. In contrast, a polymer of identical chemical composition, but consisting of random sequences of units, showed biphasic textures (which could be very coarse) over a temperature range of 120 °C. Their interpretation for this relied on the concept of *polyflexibility* – the occurrence of a range of persistence lengths for the chain population. For chains with a perfectly regular sequence of units along the chain, persistence length will be fixed so that the system is monoflexible. Differences between the chains will only arise from chain length variations, i.e. polydispersity. This will not be the case for chemically disordered chains. Their calculations showed that polyflexibility could lead to phase separation occurring over a large temperature range, whereas a sharp liquid crystal–isotropic transition was predicted for monoflexible chains.

Fig. 6.19 Spherulites suspended in an isotropic matrix – PBLG in dichloromethane. (Courtesy of A. P. Ritter.)

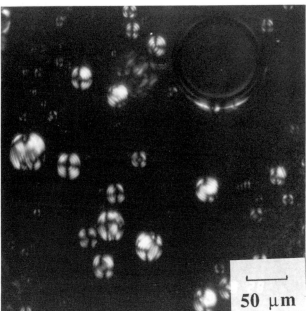

However, it should be noted that, depending on the precise chemistry, chemical disorder need not invariably lead to polyflexibility.

6.6   Molecular weight effects

The role of molecular weight in determining textures has been referred to at several points in the preceding sections. In the case of side-chain polymers, where the pendant groups are effectively decoupled from the backbone and the ordering of the side-chain mesogenic units determines the liquid crystallinity, molecular weight may in fact be less crucial. This view is supported by viscosity data (Chapter 7.2.2): the overall length determines the viscosity and hence the time required to approach an equilibrium structure, but the elastic constants, which will determine the nature of the structure, are controlled by the length of the flexible spacer, i.e. how effectively the mesogenic units are decoupled. It is thus not surprising that side-chain liquid crystalline polymers show textures very reminiscent of their small-molecule counterparts. For main-chain polymers with flexible spacers which show smectic phases, the layer spacing is determined by the length of the repeat unit [6.38], and is comparable with small-molecule mesogen dimensions. Again the similarity with small-molecule textures is perhaps to be expected.

Main-chain rigid polymers, however, which only appear to form nematic phases can in no sense be considered similar to a small-molecule liquid crystal, and their behaviour must be regarded as distinct. Nevertheless, the geometrical requirements of the nematic phase and its defects (disclinations) will still be common to both small-molecule and polymeric liquid crystals, but the molecular weight factor will be crucial in determining exactly which defects are energetically favoured.

6.7   Domains

So far in this chapter, particular well-developed textures and defects have been discussed, but many polymers (particularly thermotropic nematics) show none of these, having only a fine grained anisotropic texture with an apparently random distribution of lines of extinction. Upon rotation of crossed polars the overall appearance remains unchanged, although the position of any given dark line is likely to change smoothly. This kind of structure has been termed a Schlieren texture (e.g. [6.16]) but it is on a much finer scale than the typical Schlieren structure of small-molecule liquid crystals, and it may not be possible to identify individual disclinations (Fig. 5.7). The scale of the structure is typically only a few microns and it is often impossible (unlike in small-molecule liquid crystals) to remove this texture to give macroscopic regions of constant orientation.

Loosely speaking, this structure has been referred to as a domain structure. Indeed the term domain is frequently used in the arena of liquid crystalline polymers to describe a region within which the orientation is essentially uniform, and distinct from neighbouring domains which themselves possess a uniform orientation but whose director is not parallel to the first domain. Domains have also been invoked to describe rheological properties (see Chapter 7) (although in this case one definition of a domain was given as a region over which the averaged orientation was zero which therefore implies a different meaning to the word [6.39, 6.40]). The assumption is that each domain is a discrete entity with a well-defined boundary.

However, there is no concrete evidence to identify the boundary as representing a discontinuity in the director field. On the contrary it seems from analysis based on the high resolution technique of transmission electron microscopy that in some cases the director varies smoothly and continuously [6.40] whereas in other cases regions of essentially uniform orientation are separated from neighbouring regions by a narrow 'wall' in which the molecular orientation changes rapidly, but nevertheless smoothly [6.41].

Some considerable discussion of the nature of domains can be found in the Discussion Meeting of the Chemical Society on liquid crystalline polymers [6.42]. To some extent it is a question of definition, and a useful working hypothesis may be that domains are regions possessing a uniform orientation over dimensions many times larger than the boundary which separates neighbouring domains; within the boundary the orientation changes rapidly albeit smoothly. Further studies are needed before it can be seen whether this is a useful definition.

7 Induced ordering; response to applied fields and processing

7.1 Surface induced alignment

Alignment produced by surfaces has already been touched upon at various places in the preceding chapters, and only a few words will be added here. Whereas, for small-molecule liquid crystals, several techniques are now available for developing uniform (homogeneous or homeotropic) orientation at surfaces, their application to orientation in liquid crystalline polymers is proving more difficult to achieve. This has complicated attempts to measure, for instance, the elastic constants via the Fredericks transition.

For PBLG it does seem possible to obtain uniform orientations. For instance, in dioxane solutions it adopts a homeotropic alignment at a glass surface. Parallel alignment has been achieved at a Teflon surface, and also seems to be adopted at a free (air) surface [7.1]. Because of this difference in orientation at the two surfaces, a thin layer of PBLG solution on a glass substrate has a continuously deformed director through the thickness of the layer, as has been documented by Meyer [7.1]. This distortion leads to splay energy, which can be compensated by a splay deformation of opposite sign in the plane of the sample; the total splay now approaches zero. This deformation cannot be continuous but breaks up into domains separated by disclinations. Figures 7.1(a) and (b) show the director orientations in two perpendicular planes. It is because splay is the distortion with the highest energy (see Chapter 6) that 'splay compensation' is particularly favoured.

A rather similar effect, on a finer scale, has been observed by transmission electron microscopy in a thermotropic random co-polyester of HBA and HNA ⟨32⟩. In this case the use of a rocksalt substrate seems to encourage the homeotropic orientation, although if the rocksalt is coated with a thin layer of carbon the tendency towards perpendicular alignment is suppressed [7.2]. Thus, as expected, orientation at the surface is chemically specific. A second factor which will control surface orientation is the molecular weight of the liquid crystalline polymer. Since for homeotropic alignment one end of the chain must be situated at the surface, such alignment will be most readily achieved by short chains with their relatively high density of chain ends. This expectation is also borne out by experiments on ⟨32⟩, which show furthermore that the short chains appear to segregate in highly localized regions of near-homeotropy.

7.2 Orientation under flow
7.2.1 *Introduction*

One of the characteristics of liquid crystals, both small molecule and polymeric, is their low viscosity compared with isotropic fluids. This low value arises from the ready alignment of the molecules under flow, since local alignment already obtains. However, the *rheology* of liquid crystals is complex; as a consequence of their anisotropy, a single viscosity coefficient is not sufficient for a complete

Fig. 7.1 Splay–bend distortions in a polymer layer with homeotropic orientation at the lower (glass) surface and planar orientation at the upper (air) surface: (*a*) vertical cross section and (*b*) planar view. Neighbouring domains are separated by disclinations. (After [7.1].)

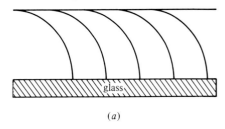

(*a*)

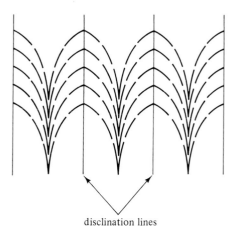

disclination lines

(*b*)

description. The formalism necessary for a full rheological description will be briefly discussed in the next section, but comparatively little detailed experimental work has been done for liquid crystalline polymers which enables comparisons to be made.

Just as PBLG was one of the first liquid crystalline polymer systems to be studied morphologically, so also was it the first to be studied rheologically. Hermans, in 1962, measured its viscosity as a function of concentration, molecular weight and shear rate [7.3], and discovered that the viscosity in the anisotropic phase is significantly lower than in the disordered isotropic (lower concentration) liquid, in line with

Fig. 7.2 Viscosity versus concentration for PBLG of molecular weight 270 000. (After [7.3].) Solvent is *m*-cresol.

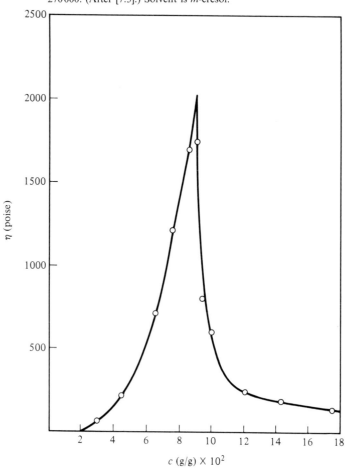

c (g/g) $\times 10^2$

earlier studies of small-molecule materials such as *p*-azoxyanisole (PAA) $\langle 2 \rangle$. Figure 7.2 shows that the viscosity versus concentration curve exhibits a predominant maximum. The position of this maximum was initially identified with v'_p, the critical concentration for the onset of liquid crystallinity in the quiescent state, but it is now thought that the anisotropic phase starts to form at concentrations somewhat lower than the maximum in the viscosity curve.

Hermans postulated that, whereas at low shear rates the molecules in an isotropic solution remain unoriented, in the anisotropic solution the director readily aligns with the shear direction, and the viscosity drops. Thus, as indicated above, the reduction in viscosity is a manifestation of the ease with which an anisotropic solution can orient under flow.

These initial findings on PBLG have since been reproduced on a wide range of both lyotropic and thermotropic polymers (see e.g. reviews by Baird [7.4] and Wissbrun [7.5]). In many respects the rheology of liquid crystalline solutions differs from isotropic solutions. Liquid crystalline solutions often exhibit a yield stress followed by non-Newtonian behaviour, i.e. a viscosity which depends on shear rate (this latter behaviour may also be exhibited by conventional polymer solutions). The viscosity of high molecular weight flexible polymers is known to depend on molecular weight to the 3.4 power, but a much stronger molecular weight dependence is observed for lyotropic rigid rod macromolecules in dilute solution. Finally, when measurements beyond simple viscosity data are collected, another curious difference between conventional polymer solutions (and melts) and liquid crystalline phases is observed. For certain combinations of concentration and shear rate, a negative first normal stress difference has been measured. This is the difference between the normal (not shear) stresses along and perpendicular to the stream lines of a flowing fluid. This behaviour has been seen in liquid crystalline solutions of PBLG and poly(ε-carbobenzoxy-L-lysine) in *m*-cresol, and also in thermotropic copolyester melts. Although it is not easy to produce a physical explanation of the negative first normal stress difference, it does emerge as a consequence of the Ericksen–Leslie continuum theory described below.

Detailed rheological measurements on liquid crystalline polymers are still in their infancy, and it must additionally be borne in mind that, as with their small-molecule counterparts, three *Miesowicz viscosity coefficients* are required to describe the complete rheological response. Furthermore, this response may depend on prior history, probably because of the presence of defects and/or domains and 'crystallites'. However, it should be noted that the lower viscosities of the liquid crystalline phases, and the 'shear thinning' of solutions or melts (i.e.

the lowering in viscosity with an increase in shear rate or stress) is of considerable importance in commercial applications. For example, thermotropic polymers can be moulded at relatively low pressures while achieving excellent filling of moulds with detailed and tortuous shapes.

The Ericksen–Leslie model, a continuum model (outlined in Section 7.2.2), is not the only model to describe the flow of rigid-rod systems. Doi [7.6] has extended molecular models for dilute solutions of rigid rods to concentrations above the critical value necessary for liquid crystallinity. These molecular models will be described in Section 7.2.3. However, a fundamental problem remains with both types of theoretical approach: liquid crystalline polymer melts and solutions are rarely uniform, but consist of domains as revealed for example by the pioneering rheo-optical studies of Asada *et al.* [7.7]. Phenomenological models to incorporate these observations have been attempted by various workers, but as yet a clear picture has not emerged. It does however seem apparent that there is an intimate relationship between textures, as can be observed by polarizing microscopy, and the measured rheological parameters. Consequently it is especially important that rheological measurements are made in conjunction with structural studies.

As will be apparent from the experimental results presented below, the vast majority of rheological data collected to date have been on main-chain systems, with very little information available on side-chain polymers. This is partly because of the different applications envisaged for main- and side-chain polymers. However, a rheological understanding of side-chain polymers will become increasingly relevant to processing control in device fabrication.

7.2.2   *The Ericksen–Leslie (continuum) theory for flow of nematics ('nematodynamics')*

A fundamental property of nematic solutions under flow is that there is a coupling between the velocity field $\mathbf{v}(\mathbf{r})$ and the director $\mathbf{n}(\mathbf{r})$. Thus in general, flow will perturb the alignment and, conversely, changes in alignment will lead to flow. Ericksen [7.8] was the first to tackle the hydrodynamics of the coupled velocity and director fields followed by Leslie [7.9] and the theory is usually known by their combined names, sometimes also coupled with that of Parodi who made an additional simplification. Derivation of the Ericksen–Leslie equations is beyond the scope of this text, and the interested reader is referred either to the original papers, or to other texts, e.g. [7.10].

However, the final equations are discussed below, and used as a basis for understanding the salient issues.

Different schools have used different notations in the basic equations. In this text we shall confine ourselves to a single formulation. The basic equation can be written as:

$$\boldsymbol{\sigma} = \alpha_1(\mathbf{n}\cdot\mathbf{A}\cdot\mathbf{n})\mathbf{nn} + \alpha_2\,\mathbf{nN} + \alpha_3\,\mathbf{Nn} + \alpha_4\,\mathbf{A} + \alpha_5\,\mathbf{nn}\cdot\mathbf{A} + \alpha_6\,\mathbf{A}\cdot\mathbf{nn} \tag{7.1a}$$

(the terms will be discussed below). It is often convenient to use suffix notation with the convention of summing over repeated indices, in which case equation (7.1a) can be rewritten as

$$\sigma_{ij} = \alpha_1(n_k A_{kl} n_l) n_i n_j + \alpha_2 n_i N_j + \alpha_3 N_i n_j + \alpha_4 A_{ij} + \alpha_5 n_i n_k A_{kj} + \alpha_6 A_{ik} n_k n_j \tag{7.1b}$$

From the suffix notation it is clear that $\boldsymbol{\sigma}$ and $\mathbf{A}$ are tensors, as they have two subscripts, whereas $\mathbf{n}$ (the familiar director) and $\mathbf{N}$ are vectors having a single subscript. The six coefficients α_1 to α_6 (which may be positive or negative) are known as the *Leslie coefficients*, and have the units of viscosity. The tensor σ_{ij} represents the viscous shear stress in the nematic fluid. As shown in equations (7.1), it is the sum of six terms expressing the torques developed in the anisotropic liquid, and these are associated with both rotational and non-rotational flow.

If an external field is present, such as a magnetic field $\mathbf{H}$, a second equation is required:

$$H_i = (\alpha_3 - \alpha_2)N_i + (\alpha_3 + \alpha_2)A_{ik} n_k \tag{7.2}$$

where $\mathbf{n}\times\mathbf{H}$ is the torque per unit volume caused by the external field. (This torque arises because the field will have an orienting effect on the director independent of that generated by the flow.)

In addition, the Ericksen–Leslie theory assumes that the fluid is incompressible and that the magnitude of the director is fixed. It can also be shown that only five of the six Leslie coefficients are independent, so that

$$\alpha_3 + \alpha_2 = \alpha_6 - \alpha_5 \tag{7.3}$$

a relationship often referred to by the name of its discoverer, Parodi.

The meaning of the symbols in the above equations will now be discussed in turn.

(a) $\boldsymbol{\sigma}$: In suffix notation this is written as σ_{ij}. It represents the shear stress associated with the viscous flow. In the case where i and *j* are orthogonal, *j* represents the direction of the shearing force and *i* the direction of the normal to the planes being sheared. Thus for forced shear of a nematic between two plates, *j* would be parallel to the shear direction, and *i* would be normal to the plates.

(b) **n**: The unit vector representing the director of the nematic.

(c) **A**: This is a shear strain rate tensor, describing only the change in shape associated with the flow, and not the rotation. It is a symmetric tensor which can be expressed in terms of gradients in the velocity field:

$$A_{ij} = \frac{1}{2}\left(\frac{\partial v_i}{\partial x_j} + \frac{\partial v_j}{\partial x_i}\right) \tag{7.4}$$

where **v** is a linear velocity with components v_i. It is interesting to note that for an isotropic liquid, equation (7.1b) reduces to

$$\sigma_{ij} = \alpha_4 A_{ij} \tag{7.5}$$

α_4 being in this case equivalent to 'the' viscosity η, as commonly used.

(d) **N**: This is a vector expressing the rate at which the director orientation changes with time with respect to the background fluid. It can be written as:

$$\mathbf{N} = \frac{d\mathbf{n}}{dt} - (\boldsymbol{\omega} \times \mathbf{n}) \tag{7.6a}$$

with $\boldsymbol{\omega} = \frac{1}{2}\mathbf{curl\ v}$ (7.6b)

(**curl v** is called the vorticity of the fluid). In component form this can be written:

$$\boldsymbol{\omega} = \frac{1}{2}\left(\frac{\partial v_y}{\partial z} - \frac{\partial v_z}{\partial y}, \frac{\partial v_z}{\partial x} - \frac{\partial v_x}{\partial z}, \frac{\partial v_x}{\partial y} - \frac{\partial v_y}{\partial x}\right) \tag{7.6c}$$

In suffix notation equation (7.6a) can then be written:

$$N_i = \frac{dn_i}{dt} - \frac{1}{2}\left(\frac{\partial v_i}{\partial x_k} - \frac{\partial v_k}{\partial x_i}\right)n_k \tag{7.7}$$

with summation implied over the repeated index *k*.

Thus equations (7.6b) and (7.6c) represent that element of rotation associated with the shear strain rate $\partial v_i/\partial x_j$ which is not included in **A**. The total change in shape of an initially square element of fluid can therefore be depicted schematically as the sum of the two components, shape change and rotation, as shown in Fig. 7.3.

From these descriptions it can be seen that the terms involving α_2 and α_3 in equations (7.1) are associated with rotational flows and the terms involving α_5 and α_6 with non-rotational flows (no dependence on the vorticity ω). It can be seen from the suffix notation representation in equation (7.1b) that the four terms involving α_2, α_3, α_5 and α_6 are equivalent to the different ways the director field **n** can couple to the flow fields. The first term (unlike these terms) is symmetric and describes the stretching that can be produced by a non-rotational flow.

The three main (Miesowicz) viscosities, η_a, η_b and η_c, describe the viscosity measured in a simple shear experiment for the three geometries shown in Fig. 7.4 in which **n** (imposed, for instance, by a strong magnetic field **B**) is either parallel or orthogonal to the velocity gradient. The three viscosities η_a, η_b and η_c (a, b and c relating to the

Fig. 7.3. Any deformation can be represented as having two components, one associated with a shape change (*a*), the other with a rotation (*b*). The net sum of these two converts a square into a rotated rhombus (*c*).

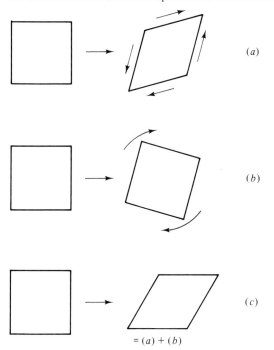

(a)

(b)

(c)

$= (a) + (b)$

three geometries shown in Fig. 7.4) can be related to the α coefficients by

$$\eta_a = \tfrac{1}{2}\alpha_4 \tag{7.8a}$$

$$\eta_b = \frac{\alpha_4 + \alpha_3 + \alpha_6}{2} \tag{7.8b}$$

$$\eta_c = \frac{\alpha_4 + \alpha_5 - \alpha_2}{2} \tag{7.8c}$$

Measurement of these viscosities is difficult because the presence of surfaces may in itself lead to perturbations in alignment, as indeed will the flow. Thus to obtain the uniform orientation required is not a trivial matter. Additionally all defects (disclinations) must be eliminated. Experiments on small-molecule liquid crystals along these lines were first carried out by Miesowicz on PAA [7.11].

Apart from the Miesowicz viscosities, it is also useful to use viscosities that specifically relate to splay (subscript S), twist (subscript T) and bend (subscript B) distortions. These viscosities can be related to the Miesowicz viscosities and the Leslie coefficients as follows: (see [7.12] for example).

$$\eta_T = \alpha_3 - \alpha_2 \tag{7.8d}$$

$$\eta_S = \eta_T - (\eta_c - \eta_b - \eta_T)^2 / 4\eta_b \tag{7.8e}$$

$$\eta_B = \eta_T - (\eta_c - \eta_b + \eta_T)^2 / 4\eta_c \tag{7.8f}$$

η_T can be measured from the dynamics of the Fredericks transition [7.13] by measuring the characteristic time for relaxation τ when the

Fig. 7.4 The geometry required to measure the three Miesowicz viscosities η_a, η_b and η_c: (a) the director is perpendicular to the plane containing the velocity gradient; (b) the director is parallel to the velocity and (c) the director is perpendicular to the velocity but in the same plane as the velocity gradient. The uniform orientation is commonly achieved by the use of a magnetic field.

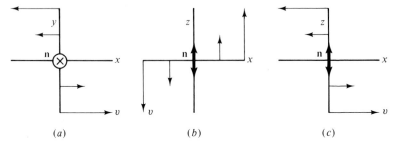

(a)   (b)   (c)

field is decreased to some value below the critical field. An alternative technique involving reorientation in the strong magnetic field inside an NMR apparatus is based on looking at the separation of the doublet in the NMR signal that arises from the anisotropy of the liquid crystalline phase. By measuring this separation (which is a measure of the orientation function) as a function of time after an aligned sample is rotated in the field, η_T, α_1, α_2, and η_c can be determined, together with the ratio K_3/K_1 [7.14]. η_T can also be found from photon correlation spectroscopy (otherwise known as dynamic light scattering), when strictly a ratio of an elastic constant to a viscosity is measured. This technique has been used for a racemic mixture of poly(γ-benzyl-glutamates) (PBG) [7.15] and for PBTZ [7.12, 7.16]. A discussion of the data is presented in Section 7.2.4.

The method involving the dynamics of the Fredericks transition for measuring η_T has been used for a family of siloxane side-chain polymers [7.17]. In this case the backbone itself must play a part, and so the hydrodynamic behaviour resembles a normal flexible polymer melt, and η_T is found to be almost linearly dependent on the backbone length. The temperature dependence of the rheological behaviour of carbon backbone side-chain polymers has also been studied [7.18]. For one polymer a maximum was found in the melt viscosity in the vicinity of the liquid crystalline to isotropic transition. The determined α_2 and α_3 values correlated well with the rheological determination of melt viscosity. Knowledge of the viscosity coefficients is particularly important in understanding the instabilities that may arise when, for example, a strong magnetic field is applied (see the discussion of the Fredericks transition in Section 6.2.3 and Section 7.5).

7.2.3 *A molecular model for flow in liquid crystalline polymers*

The Ericksen–Leslie theory has recently been applied to liquid crystalline polymers, although, as it is essentially a continuum model derived for small-molecule mesophases, it is not able to predict viscoelasticity in polymers. On the other hand the model due to Doi and Edwards is a molecular model, developed explicitly for solutions of rigid-rod macromolecules. A detailed discussion of the theoretical basis for the work can be found in the book by Doi and Edwards [7.19].

The essence of the model can be expressed as follows. For a solution of rods, of length L, diameter d and number per unit volume, c, three regimes can be identified (c is related to the volume fraction v_p by $v_p = (c\pi d^2 L)/4$). For concentrations where $c \ll 1/L^3$ the solution is dilute and there are essentially no interactions between the rods. The solution is said to be unentangled since each rod feels no constraint to its motion due to the other rods. At the other extreme when $c \approx 1/L^2 d$, the Onsager condition for the onset of liquid crystallinity is met, and spontaneous alignment occurs. In the intermediate range where $1/L^3 < c < 1/L^2 d$, the solution is *entangled* and the free rotation of each rod is severely impeded by other rods although the system remains thermodynamically ideal.

The entanglement effect, equivalent to the condition that two rods cannot pass through each other (and in that sense at least equivalent to the concept of an entanglement for a flexible polymer chain), can arise even for rods of zero thickness. It can therefore be distinguished from the *volume exclusion* effect, which also gives rise to an effective repulsive potential but will vanish as d approaches zero.

Once the rods are entangled, their rotational motion about any axis normal to their length becomes severely restricted, as does translation perpendicular to their length. Translational motion along the length of each rod will be essentially unrestricted and can still be represented by D_{t0}, the free translation diffusion coefficient for a rod

$$D_{t0} = \frac{kT \ln (L/d)}{2\pi\eta_s L} \tag{7.9}$$

where η_s is the solvent viscosity and k Boltzmann's constant. On the other hand, since in a direction perpendicular to any rod its neighbour is only some small distance a_c away, translational diffusion in this direction can be neglected.

In the model, as originally envisaged by Doi and Edwards, rotational motion of a rod (the 'test rod') was restricted to some small solid angle $\Delta\Omega \approx (a_c/L)^2$ until a neighbouring rod had diffused a distance L, thus removing a 'bar' of the 'cage' within which the test rod was trapped. Within this scheme the rotational diffusion constant D_r can be shown to be

$$D_r \approx \left(\frac{D_{t0}}{L^2}\right)(cL^3)^{-2} \approx \frac{D_{r0}}{c^2 L^6} \tag{7.10}$$

where

$$D_{r0} \approx \frac{D_{t0}}{L^2} \approx \frac{kT}{\eta_s}$$

D_{r0} is the free rotational diffusion constant in dilute solution. The requirement for the neighbouring rod to diffuse a distance L has recently been challenged as being too restrictive and experiments seem to support this view. However, a rapid decrease in D_r is predicted as the rod length L increases, even if the precise details remain uncertain.

Experiments suggest that the Doi–Edwards model for semi-dilute

solutions is qualitatively correct. However, results from both birefringence measurements in extensional flow fields, and dynamic light
scattering measurements of the rotational diffusion coefficient indicate
that its reduction from the value for a free rod, D_{r0}, occurs at
significantly higher concentrations than those predicted. If one
expresses equation (7.10) as

$$D_r \approx \frac{\beta D_{r0}}{c^2 L^6}$$

where the constant $\beta \approx 1$ in the original model, experiments suggest
that a much larger value of $\beta(\approx 1000)$ is appropriate, supporting the
view that neighbouring rods need not diffuse along their total length to
release the test rod.

The dramatic decrease in D_r as c and L increase, has a profound
effect on the rheological properties of the solution. In dilute solutions
of rod-like molecules the viscosity can be written as

$$\eta = \eta_s + \frac{ckT}{D_{r0}}$$

and since $D_{r0} \approx \frac{kT}{\eta_s L^3}$ (see [7.19]), then

$$\eta \approx \eta_s(1 + cL^3) \tag{7.11}$$

Equation (7.11) shows that for concentrations, $c \ll 1/L^3$, the viscosity
is little changed from the pure solvent. However, once $c \gg 1/L^3$, D_{r0}
must be replaced by D_r in equation (7.11), and hence using equation
(7.10)

$$\eta = \eta_s(1 + (cL^3)^3) \tag{7.12}$$

The viscosity is dramatically increased, with a strong molecular weight
dependence. This conclusion is supported by experimental work,
although the precise numerical value of the power law remains unclear.
Furthermore, since the viscosity is largely associated with rotational
degrees of freedom, it is very sensitive to both shear rate and shear
history, giving rise to a strong non-Newtonian viscoelasticity, as seen
experimentally.

But what about solutions sufficiently concentrated that a nematic
phase has formed? The original Doi–Edwards model explicitly
excluded this regime, but Doi has more recently considered the case
where $c \geqslant 1/dL^2$. A full discussion of this treatment can be found in
[7.19] and only an outline will be given here. The approach builds on

the formulation of the nematic–isotropic transition due to Onsager (e.g. [1.12]). The free energy $\mathscr{A}$ of the solution can be expressed in terms of the order parameter S as

$$\mathscr{A}(S, v_{\mathrm{p}}) = \frac{1}{2}\left(1 - \frac{v_{\mathrm{p}}}{v_{\mathrm{p}}''}\right)S^2 - \frac{v_{\mathrm{p}}}{3v_{\mathrm{p}}''}S^3 + \frac{v_{\mathrm{p}}}{2v_{\mathrm{p}}''}S^4 \qquad (7.13)$$

where v_{p} is the volume fraction of rods, and v_{p}'' the concentration above which the isotropic phase is unstable thermodynamically. Figure 7.5 shows the form of S as a function of concentration – for a given value of $v_{\mathrm{p}}/v_{\mathrm{p}}''$ the solution for S is that which corresponds to a minimum in $\mathscr{A}(S, v_{\mathrm{p}})$. It is interesting to note that between $v_{\mathrm{p}} = (\frac{8}{9})v_{\mathrm{p}}''$ and $v_{\mathrm{p}} = v_{\mathrm{p}}''$, both isotropic ($S = 0$) and nematic states can coexist. In this regime the order parameter of the nematic is given by:

$$S = \frac{1}{4} + \frac{3}{4}\left(1 - \frac{8v_{\mathrm{p}}''}{9v_{\mathrm{p}}}\right)^{\frac{1}{2}} \quad \text{for} \quad \frac{8v_{\mathrm{p}}''}{9} < v_{\mathrm{p}} < v_{\mathrm{p}}'' \qquad (7.14)$$

As the rods become more aligned and S increases, the motion of a 'test rod' will become less hindered by the surrounding cage of rods. It is this effect that leads to this reduction in viscosity with increasing orientational order. (This is equivalent to the observation that it is easier to extract a pencil from an orderly aligned pile than from a random heap.) It can then be shown [7.19, 7.20] that the viscosity can be written in terms of S:

$$\frac{\eta}{\eta^*} = \left(\frac{v_{\mathrm{p}}}{v_{\mathrm{p}}''}\right)^3 \frac{(1-S)^4(1+S)^2(1+2S)(1+3S/2)}{(1+S/2)^2} \qquad (7.15)$$

Fig. 7.5 The order parameter S as a function of the reduced variable $v_{\mathrm{p}}/v_{\mathrm{p}}''$, where v_{p} is the volume fraction and v_{p}'' the critical volume fraction above which the isotropic phase is thermodynamically unstable. (After [7.20].)

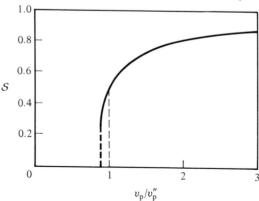

where η^* is the viscosity of the fluid at v_p''. This curve, which is plotted in Fig. 7.6, is a universal curve when plotted against v_p/v_p'' and shows, in agreement with experimental results (Fig. 7.2), a strong maximum in the vicinity of the phase transition. Note that when $S = 0$, $\eta/\eta^* = (v_p/v_p'')^3$, but as $S \to 1$ in the mesophase $\eta/\eta^* \to 0$.

7.2.4 *Rheological data and comparison with theory*

In carrying out rheological experiments for liquid crystalline polymers, two approaches can be pursued. One can either use standard rheological equipment, such as may be found in a conventional polymer laboratory, and measure (typically) a dynamic viscosity, and storage and loss moduli, G' and G'' respectively, or one can attempt to measure the various viscosities such as $\eta_{a,b,c}$ and $\eta_{\mathrm{T,S,B}}$ which can be more readily related to theory. These measurements have been achieved by measurements of birefringence to give the order parameter, S, and static and dynamic light scattering using suitably aligned samples.

The ideas behind the use of light scattering were first put forward for small-molecule liquid crystals (see e.g. [7.10]). The intensity of light

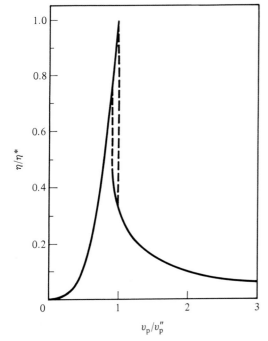

Fig. 7.6 Steady state viscosity at zero shear rate plotted as a function of the reduced variable v_p/v_p'', where v_p is the volume fraction and v_p'' the critical volume fraction above which the isotropic phase is thermodynamically unstable. (After [7.20].)

scattered in static light scattering (as measured for instance by the Rayleigh ratio), can be related to the three Frank constants, and the correlation times determined from dynamic light scattering can be related to a combination of the elastic constants and the viscosities. By collecting data over a range of scattering angles and using different polarizations, it is possible to evaluate the various constants. However, because of the difficulties associated with precise quantification of intensities in light scattering, it is more usual to present data in the form of ratios of the various parameters. A useful summary of the procedure can be found in [7.21]. Alternatively the parameters can be determined by first determining the order parameter. To carry out the measurements, uniform orientation is set up and birefringence measurements are used to obtain the order parameter. From this value there are theoretically predicted relationships between the six viscosity coefficients and S. Dynamic light scattering measurements are then used to measure the elastic constants. It should be noted that there are two different derivations for obtaining the α coefficients from S and these differ most crucially with regard to the sign of α_3. Furthermore, a single value of S does not seem able to fit all the α values determined from light scattering. Thus although there is only a very limited amount of data available so far, the validity of the theoretical relationships between S and the α_i has not yet been convincingly demonstrated.

Data gathered using both rheological and light scattering techniques will be presented briefly here. Both have their advantages, the latter providing more fundamental information but the former being more relevant to processing situations where particular orientation regimes are not specifically imposed. In practice, nearly all the data for thermotropic systems have been gathered using standard rheological techniques, whereas the light scattering experiments are more suitable for lyotropic solutions which can be obtained as samples of uniform orientation with greater ease.

One of the major problems associated with comparing theory and experiments in these systems is the lack of ideality of any actual polymeric solution. No system is ever composed of perfectly rigid, uniform (monodisperse) infinitely thin rods of length L. Polydispersity and chain flexibility inevitably modify the results in ways which are almost impossible to assess. Another complication is that a solution may not, as assumed in the Doi–Edwards model, be thermodynamically ideal. In this case an apparent increase in the diffusion coefficient with concentration may be measured because of a repulsive thermodynamic interaction which more than compensates for the decrease in the D_r appropriate to a θ (thermodynamically ideal) solvent. The problems are typified by work on **PBLG** where there is still some disagreement as to the shear rate at which shear-thinning commences, or even whether it occurs first in isotropic or liquid crystalline solutions. One complication is that flow may not only induce alignment, as we have seen above, but it may actually modify the concentration at which the liquid crystalline phase forms [7.22].

The challenge of designing rheological experiments to test theory is intensified by the difficulty of obtaining monodomain samples in high molecular weight liquid crystalline polymer systems. There is also the complicating factor that lyotropic systems, and to some extent thermotropic ones, show biphasic behaviour in the region of the liquid crystalline to isotropic transition. Furthermore, where random co-polymers have been synthesized as a method of reducing crystalline melting points in thermotropic systems, there is the possibility that a low volume fraction of crystallites will exist in the melt, particularly at temperatures not far above the melting point. These factors can lead to marked thermal history effects, and affect the temperature dependence of viscosity as the crystals melt out with increasing temperature. The complexity of a system in which both biphasic regions and partially melted crystals occur has been emphasized by the experiments of Wunder *et al.* [7.23]. The apparent viscosity exhibited two maxima, one corresponding to the presence of partially melted crystals, the other to the coexistence of anisotropic and isotropic phases. In the vicinity of both these maxima there were marked deviations from Newtonian behaviour.

Domains are undoubtedly present in both lyotropic and thermo-tropic systems, but exactly how they are modified by the nature of the velocity field is still uncertain. However, it is clear that the structure of the domains changes as shear proceeds, and that at high rates of shear the domains may disappear completely. The situation has been portrayed as an initial polydomain structure, which breaks up into smaller domains during shear [7.24]. More recently the domain structure has been related to disclinations. Two approaches have been used. Alderman and Mackley [7.25] suggested that at some critical shear, multiplication of disclinations occurred, followed by further increase in their density, and a corresponding decrease in size, as shear continued to increase. Ultimately their size is assumed to decrease beyond the resolution of the light microscope so that it appears from the birefringence as if uniform orientation has been achieved. Modelling of this disclination increase and its effect on both the optical appearance and the measured rheological parameters has been attempted [7.26]. In particular this model has explored the possibility that the measured sharp decrease in G' and G'' above $\sim 10\%$ strain in a random copolyester $\langle 32 \rangle$ (shown in Fig. 7.7) can be identified with the change in texture due to disclination multiplication. (Similar fall-offs in the measured values of both the dynamic viscosity and the moduli have been measured in other polymers.) On the other hand, Kulichikhin [7.28] considers the appearance of uniform birefringence at high strains to correspond to the destruction of disclinations, which he likens to cross-links. He thus perceives the low shear state to have high viscosity because of the presence of disclinations, and the viscosity to drop as the disclination density falls. Thus a crucial question which needs to be answered to resolve these different viewpoints is the nature of the structure of the system at high shears – is it uniformly oriented or not?

Phenomenological models have been developed to fit experimental rheological data for main-chain polymers. For certain liquid crystalline polymers, when viscosity versus shear rate curves are plotted, three distinct regions can be identified, as shown schematically in Fig. 7.8. Regions I and III are shear-thinning (i.e. the viscosity decreases with increasing shear rate), and in region II there is a plateau viscosity. Other compositions show only parts of this curve, and if the portion exhibited is shear-thinning it is often impossible to decide whether this should be interpreted as region I or III. Note that this means in region I the viscosity is increasing (apparently without limit) as the shear rate

Fig. 7.7 Moduli $G'(\triangle)$ and $G''(\square)$ as a function of strain (in %) for $\langle 32 \rangle$ at a strain rate of 10 rad s^{-1} and 320 °C. (After [7.26].)

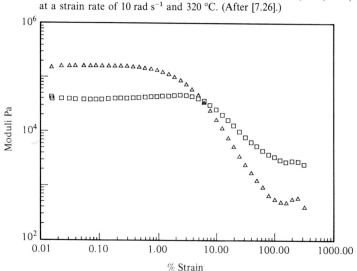

Fig. 7.8 The three regions of behaviour when viscosity is plotted versus shear rate. Regions I and III are both shear thinning, but there is no shear rate dependence in region II. (After [7.24].)

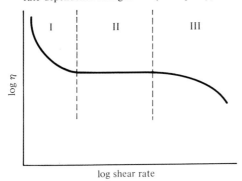

is reduced. In contrast to the shear-thinning behaviour of regions I and III, in the plateau region II, the viscosity seems insensitive to shear rate. It is in this region that viscosity–concentration data, such as shown in Fig. 7.2 have been collected, but there is no rigorous explanation for the lack of shear rate sensitivity. Overall, the indications are that no single explanation will suffice for all the systems studied, and that the results are likely to depend critically on sample history and micro-structure.

Turning now to lyotropic systems for which light scattering has been used to measure both the Frank constants and the viscosities, two different polymers have been studied – PBT and racemic poly(γ-benzyl-

Table 7.1

Parameter	PBG solution (after [7.15])	PBT solution (after [7.27])	Polyester known as DDA9 (after [7.14])	
			(211 °C)	(221 °C)
K_1/pN	4.1	18.2	—	—
K_2/pN	0.36	1.15	—	—
K_3/pN	4.7	8.4	—	—
η_T/Pa s	3.47	220	411	166
η_S/Pa s	3.45	190	406	152
η_B/Pa s	0.02	30	2.9	0.88
η_b/Pa s	0.016	—	3	0.89
η_c/Pa s	3.5	—	422	174
α_2/Pa s	−3.48	—	−415	−170
α_3/Pa s	−0.14	—	−4	−4

glutamate) (PBG). Table 7.1 summarizes the data obtained by Berry *et al.* [7.27] and Taratuta *et al.* [7.15]. PBT, which is the stiffer of the two molecules, gives higher values for all quantities. For both systems η_b is the lowest of the three Miesowiz viscosities, and K_2 (twist) is the lowest of the three Frank constants. For PBT the splay elastic constant is significantly higher than the bend, whereas the two are comparable for PBG.

The NMR method of Martins *et al.* has been applied to a thermotropic linear polyester of the rigid main chain type [7.14]. Data have been collected at two temperatures, and, as expected, all the coefficients are smaller in magnitude at the higher temperature. The data are included in Table 7.1. The value for η_T is approximately 10^4 higher than for a small-molecule liquid crystal nematic analogue of the polyester used, and the ratio η_b/η_c is $\sim 10^{-2}$, about two orders of magnitude lower than for a typical small-molecule liquid crystal in accord with predictions that it should scale as L^{-2} [7.1].

7.3   Processing and consequences of flow alignment
7.3.1   *Rheology*

While the emphasis so far has been on understanding rheological measurements in terms of basic flow models, either continuum or molecular, it is of greater practical importance to relate the rheological properties, as might for example be measured in a cone and plate rheometer, to the behaviour of liquid crystalline melts and solutions during processing operations such as spinning or moulding. The most significant aspects of liquid crystalline polymer flow behaviour for this purpose are:

(a)   lower viscosity as a result of liquid crystallinity, especially at high strain rates (cf. Figs. 7.2 and 7.6)

(b)   a general decrease in the viscosity with increasing strain rates. There is sometimes a Newtonian plateau at intermediate rates (Fig. 7.8)

(c)   very efficient molecular orientation in extensional strain fields.

7.3.2   *Spinning fibres*

Fibres spun from *lyotropic* solutions show good axial alignment and are the foundation of high performance products such as Du Pont's Kevlar (see Chapter 8). The fibres are spun by forcing the anisotropic solution through fine holes in a spinneret, either directly into a coagulation bath (wet spinning) or through an air gap first (dry spinning). Following solidification in the coagulation bath, the solvent diffuses out from the fibre which is then dried. The advantage of using an air gap between the spinning solution and the coagulation bath is that it provides a region of extensional flow in the fibre while it is still liquid crystalline, and thus promotes the development of good levels of orientation across the full section of the fibre. This extensional flow is known as drawdown. The air gap also permits the original solution to be held at a higher temperature, enabling more concentrated solutions and higher spinning rates to be used. The coagulation bath may not be a simple one-step process if strong acids are used (as in the production of Kevlar) which need to be neutralized without hydrolyzing the fibre. Subsequent heat treatments are often used to further improve mechanical properties. They are discussed in the next chapter in the context of mechanical properties.

For *thermotropic* fibres, the basic aim remains much the same as for lyotropics, which is to obtain the highest possible degree of chain alignment along the fibre axis and thus optimize the tensile mechanical properties. As in the case of the lyotropics, the majority of the molecular alignment is developed by extensional flow during drawdown immediately after spinning, although of course it is a liquid crystalline *melt* which is being drawn. The development of the orientation is very efficient and Calundann *et al.* [7.29] have presented data which show that virtually perfect axial alignment is achieved at very modest melt extension ratios of the order of 3. At shear rates experienced during fibre spinning, of the order of 10^3 to 10^4 s^{-1}, the viscosity of a melt of a thermotropic copolyester $\langle 32 \rangle$ lies between 10^1 and 10^2 N s m^{-2}. The marked shear thinning of liquid crystalline melts can be seen as contributing to the usefully low viscosity at these high shear rates. However, the viscosity is dependent on molecular weight, and molecular weights greater than 30000 are difficult to process into fibres [7.29]. It is also important that the melt temperature is sufficiently high to avoid the presence of residual crystallinity in the melt which has been shown to interfere with the orientation process.

It is clear that although some orientation is introduced into the surface regions of the fibre by the spinning process, subsequent drawdown is vital if the high degree of uniform alignment necessary for good axial properties is to be obtained. The spinning stage, in addition to forming the fibre itself, provides an important level of initial organization of the mesophase which has a significant influence on the final properties which can be obtained. The rheological properties of the liquid crystalline polymer solution or melt are crucial in determining the correct 'processing window'. It is necessary to obtain the right combination of temperature, molecular weight and molecular weight distribution, solvent concentration (for lyotropic processes), spinning rates, drawdown and spinneret design. If the conditions are wrong, the final properties may be compromised or instabilities develop in the process. Nevertheless, it is the inherent spinnability of liquid crystalline polymers, coupled with their extremely efficient orientation in extensional fields applied during drawdown, which make them an excellent starting point for the production of ultra high performance fibres.

7.3.3   *Extrusion and moulding*

Both these processes are only applicable to thermotropic polymers as they produce bulk samples which means that solvent extraction would be impossible at any commercially sensible rate. The geometry of *extrusion* is similar to that of fibre spinning although the bore of the die is typically in millimetres rather than microns. The shear strain rates employed are several orders of magnitude lower than for fibre spinning, being typically in the range of 1 s^{-1}. The viscosity of a thermotropic melt at these rates is not particularly low, but it will exhibit strong shear thinning. The result is a tendency for 'plug flow' in which the shear through the die is concentrated in a thin region near to the surface of the extruding shape, while the velocity of the core of the polymer at the die is comparatively uniform. The extrusion of conventional polymers produces a degree of axial molecular alignment some of which recovers as the material emerges from the die, with the result that the polymer increases in diameter. This phenomenon is known as 'die swell' and must be taken into account if the dimensions of the extrudate are required with any degree of precision, a problem which is particularly taxing if the extruded section is complex. Liquid crystalline polymers give very limited die swell on extrusion, a considerable advantage in any application requiring precision, and/or complex sections.

Injection moulding processes typically operate at strain rates an order of magnitude greater than extrusion, and are thus in the region where the lower viscosity of the liquid crystalline polymer is beginning to assert itself, and the benefits are readily apparent in excellent mould-fill characteristics. The molten polymer moves faster in the centre of the mould, whereas it is stationary at the edges where it is in contact with the walls. The result is that, at the head of the mass of polymer filling the mould, the material will be flowing from the centre towards the surfaces. This flow pattern is known as the 'fountain effect' and is not limited to thermotropic materials. However, the fact that molecular alignment in liquid crystalline polymers is especially sensitive to the flow fields means that the internal morphology of a moulding can often reflect intricate details of the flow patterns. In the case of the fountain effect, this behaviour leads to one undesirable consequence. The material at the advancing front of the polymer filling the mould will tend to be highly oriented in the direction normal to the general flow, as a consequence of the tendency of liquid crystalline polymer molecules to lie parallel to the surface as well as of the fountain flow itself. However, when two advancing slugs of polymer meet each other, as would be the case for example in moulding a wheel, the material at the point at which they join will be highly oriented normal to the general flow direction. This leads to a very weak weld line which can effectively reduce the properties of the moulding to a level much below that desired. It is therefore especially important when designing moulds for thermotropic polymers to ensure weld lines are formed at non-critical sites.

A useful review of the different processing conditions which have been studied, and which covers both lyotropic and thermotropic polymers, is to be found in [7.30].

7.3.4   Skin–core morphology

For fibres, extrudates and injection mouldings a so-called *skin–core* morphology is present. The polymer in the skin shows well-developed orientation in the direction of flow, while that in the core shows little if any. The source of this morphology can be understood easily in considering flow through a die as in Fig. 7.9. At the die the polymer at the edges is moving more slowly than that at the centre,

Fig. 7.9 Sketch showing the shear velocity gradient at an extrusion die which means that the slow moving skin material at the die accelerates to the mean velocity of the solidified bar downstream leading to elongational flow and hence well-developed alignment, whereas the core material decelerates reducing any orientation which is initially present in this region.

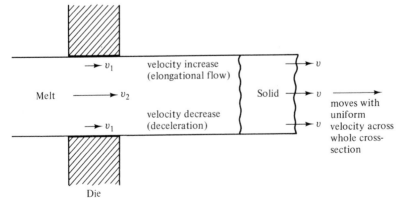

whereas downstream where the whole section has solidified the velocity will be uniform at all points. There is thus an accelerating or elongational flow in the surface regions which is highly conducive to molecular alignment, while the deceleration in the central regions will oppose what little orientation may have been introduced in this region at the die [7.24]. The argument for injection moulding geometries is not quite as straightforward in that flow patterns are more complicated. However, the result is that there will be a tensile stress in the solidifying surface material which will produce an extensional field and thus enhance orientation. In injection mouldings the skin may appear significantly shinier than the core, and in both fibres and mouldings, the skin is prone to delaminate from the core, there being a sharp boundary in between the two. An example of the peeling of the highly oriented skin of an injection moulded random copolyester ⟨32⟩ is shown in Fig. 7.10.

7.3.5   *Banded textures*

One of the more ubiquitous, if as yet imperfectly understood morphologies seen in liquid crystalline polymers is that that gives rise to the so-called *banded structure*. The phenomenon of bands was first observed by transmission electron microscopy of Kevlar fibres [7.31],

Fig. 7.10 Scanning electron micrograph of an injection moulding of a thermotropic copolyester, showing the highly oriented skin material which has begun to peel as a result of mechanical damage. (Courtesy C. M. Carr.)

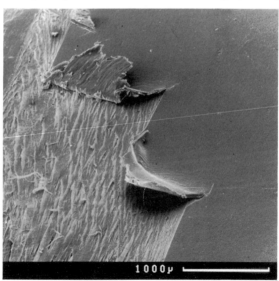

1000μ

but they have also been seen by polarized light microscopy. During the past few years the banded structure has been identified, following shear, in a very wide range of main-chain liquid crystalline polymers, both rigid and semi-flexible, thermotropic and lyotropic. Indeed, it seems to be an almost universal response if the right range of shear rate and viscosity (controlled by concentration, molecular weight and/or temperature) are selected, although since the phenomenon is not well understood, knowing a priori what the correct conditions will be is not as yet possible.

Figure 7.11 (a and b) show the appearance of bands following shear

Fig. 7.11 (a) Appearance of bands in a thermotropic polyester ⟨32⟩, as viewed in the polarizing microscope. (b) Dark field electron micrograph of bands in the same polymer; the image is formed from part of the equatorial reflection, so that only regions where the molecules are oriented correctly to scatter into this part of the diffraction pattern look bright. In each case the shear direction is vertical

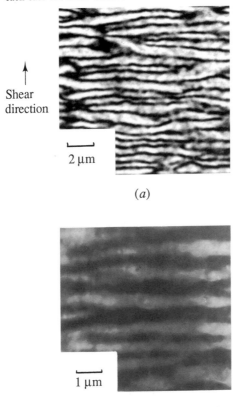

Shear
direction

2 μm

(a)

1 μm

(b)

in a thermotropic copolyester in polarized light and a dark field electron micrograph respectively. The bands always lie perpendicular to the prior shear direction. In this case the bands are observed by quenching-in the structure after shear. In general bands form after the cessation of shear, presumably as a result of some relaxation process, but there is some evidence that under some circumstances they may appear during the flow process itself.

The bands are associated with a periodic variation in director orientation about the flow axis. Initial work on Kevlar showed that in these (crystalline) fibres the bands corresponded to a 'pleated sheet'

Fig. 7.12 Schematic diagram of the molecules in the pleated sheet structure of a Kevlar fibre. (After [7.31].)

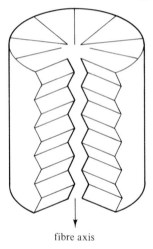

fibre axis

Fig. 7.13 The serpentine molecular trajectory underlying the banded structure in thermotropic copolyesters. Shear direction or fibre axis horizontal (After [7.32].)

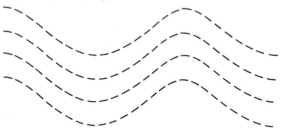

structure with radial symmetry, illustrated in Fig. 7.12, i.e., the molecular orientation took up two orientations, symmetric with respect to the fibre axis and with a sharp kink between them. Similar morphologies have been reported for PBLG ⟨22⟩, hydroxypropyl cellulose (HPC) ⟨21⟩ and a random copolyester based on molecules which themselves contain a certain proportion of sharply kinked units ⟨34⟩. On the other hand banding in a range of copolyesters ⟨32, 33⟩ etc., has been shown [7.32] to correspond to a more gentle 'serpentine' trajectory of the molecules (Fig. 7.13).

7.4   Electromagnetic field effects

7.4.1   *Electric fields*

The principles involved in the electric field orientation of liquid crystalline polymers are essentially those described for small-molecule materials in Chapter 2. The process has two components; the interaction of the applied field with permanent electric dipoles in the molecule which leads to polarization through molecular rotation, and the formation of an induced dipole which will give rise to molecular orientation by virtue of the anisotropy in its polarizability. The first component will lead to polar orientation of the molecules or groups – as long as they are not centrosymmetric – while the second will lead to quadrupole alignment whether the molecules are centrosymmetric or not. The anisotropy of the permittivity expressed as $\Delta\varepsilon = (\varepsilon_\parallel - \varepsilon_\perp)$ will lead to orientation of the molecular axes parallel to the field when positive and perpendicular to the field when negative.

Side-chain liquid crystalline polymers are the primary focus of attention with respect to behaviour in electric fields [7.33]. Not only are they being explored as analogues of conventional liquid crystals as active media in electro-optic devices, but they also open up additional possibilities of quenchability to form glasses enabling molecular orientation to be 'frozen-in' into thin unsupported films. These aspects are discussed in more detail in Chapter 8.

For any given family of side-chain polymers there are two important parameters determining the orientation behaviour, the number of units n in the backbone, and the number of decoupling units m between the backbone and the mesogenic side-chain unit, i.e. the length of the flexible spacer. The value of n is likely to affect the overall viscosity of the melt, while that of m will determine whether the mesogenic (orienting) unit is sufficiently decoupled from the main chain to be able to orient independently of it. The effect of both these parameters has been explored for various families of polymers with both silicon- and carbon-containing backbones.

To consider first the role of the flexible space length, m: it is known that without the presence of a flexible spacer a liquid crystalline phase cannot form at all, because the backbone restricts the orientation of the mesogenic side-chain units (cf. Chapter 3). Thus it is to be expected that for low m, orientation under an electric field will be hard to achieve, but it will become increasingly easy as m increases so that the active part of the polymer is able to resemble more closely a monomeric liquid crystal. This expectation is borne out by experiments on silicon backbone copolymers with ⟨65⟩ as the side group, where R is either

$$-(CH_2)_n-O-\!\!\bigcirc\!\!-\overset{\overset{\textstyle O}{\|}}{C}-O-\!\!\bigcirc\!\!-R \qquad\qquad ⟨65⟩$$

chlorine or OCH_3. The chlorine atom gives rise to a positive $\Delta\varepsilon$, the methoxy group a negative $\Delta\varepsilon$, so that by varying the ratios of the two components in a copolymer it is possible to control the dielectric susceptibility.

When the threshold voltage for alignment, V_c, was measured [7.34] (using a cell of about 10 μm thickness), it was found to increase from 4 V_{rms} (root mean square average value of voltage for AC) when $n = 6$, to 7.3 V_{rms} for $n = 5$, to greater than 50 V_{rms} for $n = 4$. Alignment could not be achieved for $n = 3$. Since the threshold voltage is related to the elastic constants, these measurements indicate the strong effect that spacer length has upon these constants.

The comparatively large value of $\varepsilon_{\|}$ in rod molecules which are asymmetric about a plane normal to their long axes, is only realized in an AC field as long as the molecule is able to reorient end-to-end at a frequency commensurate with that of the field. The frequency of this motion in a mesophase is comparatively slow, perhaps 10^3 Hz, so that at frequencies above this value there is a rapid decrease in $\varepsilon_{\|}$. As the rotational polarization which contributes to $\varepsilon_\perp$ (i.e. rotation of the rod molecule around its long axis), will have a much shorter relaxation time, possibly approaching the Debye value characteristic of small isotropic molecules, there is the possibility of a frequency range in which $\Delta\varepsilon$ is negative. This effect is the same as that well known in small-molecule liquid crystals [7.35], although the increased viscosity of the polymer melts will be reflected in lower values of the critical cross-over frequency.

Another result of the dependency of molecular reorientation rates on viscosity, is that the response time to an applied field will depend on the molecular weight of the polymer. It appears that this dependency is comparatively linear and it is not therefore very surprising that polymeric materials are not generally suitable if fast orientational response is a necessary characteristic of a device. In earlier work, little attention was paid to control of molecular weight, but without knowledge of this parameter, measurement of effects such as response time is not very meaningful, particularly when comparisons are to be made between different systems. Hence early measurements of orientation under electric fields which demonstrated rise-times $\sim$ 200 ms (comparable with small-molecule liquid crystals), may merely reflect the fact that the molecular weight was low.

The effect of electric fields on main-chain liquid crystalline polymers has been less extensively studied. For lyotropic PBLG, electric field effects seem broadly similar to magnetic field effects: the cholesteric pitch is unwound if the field strength is sufficiently large, and instabilities (see below) may be seen. Two useful reviews by Iizuka cover this area thoroughly [7.36, 7.37]. The effect of electric field orientation on the crystallization process has also been considered on the random copolyester $\langle 33 \rangle$ containing 80% hydroxybenzoic acid units [7.38]. It was found that there was a significant increase in the enthalpy of the melting endotherm, which was also shifted to higher temperatures relative to films produced in the absence of the field. The effects were attributed to dipolar interaction leading to enhanced head-to-tail coupling.

In general, electric fields are exceptionally versatile and offer control in the orientation of polymeric mesophases. Not only are they important in the area of device technology, but the measurement of AC conductivity or dielectric loss as a function of temperature provides an elegant probe to reveal characteristic motions of different components of the polymer molecule. The technique of dielectric spectroscopy is a significant one in the field, and has recently been reviewed by Haws, Clark and Attard [7.39]. However, there are difficulties associated with the use of electric fields. One particular potential problem is the migration of charged species to give a DC current which, as it involves mass transport, can also lead to disruption of the orientation in the sample. Another problem stems from the fact that, although polymer samples can withstand fields up to 5 kV mm^{-1}, the voltages necessary to obtain such field strengths in bulk samples ($>$ 1 cm thick) move one into the realm of high voltage engineering with all the attendant complications associated with breakdown. Nevertheless, it can be anticipated that there will be a continuing and increasing activity aimed at understanding and utilising electric field effects.

7.4.2   *Magnetic fields*

A molecule will orient in a magnetic field if its magnetic susceptibility is anisotropic, the anisotropy being defined by $\Delta\chi = \chi_{\parallel} - \chi_{\perp}$. The polarization is induced by the field rather than being present in the molecules as dipoles, and thus only quadrupolar alignment can result. However, magnetic fields are more readily applicable to bulk samples than electric fields, although to achieve fields between 1 and 2 tesla over a significant volume can involve magnets weighing 1000 kg. In very general terms, a magnetic field of 1 T is equivalent to an electric field of 100 V mm^{-1} in orienting power, so even if one is to consider superconducting magnets giving fields in excess of 10 T, electric fields at a practical maximum of 5 kV mm^{-1}, retain a useful advantage with respect to the inducement of orientation.

The first study of magnetic field effects on a lyotropic nematic polymer was carried out on PBA ⟨23⟩, using NMR to show that field alignment had occurred [7.40]. Subsequently it was shown that the act of filling a cell with the PBA solution itself caused significant alignment which took a considerable time to disappear under the action of a 0.7 T field perpendicular to the flow direction. The time taken for total alignment was less for a 12%, wholly anisotropic solution than for more dilute, but biphasic, solutions. Even so, some surface alignment always remained which encouraged a return to the original texture when the field was subsequently switched off.

More recently, attention has turned to thermotropic main-chain polymers, particularly those with flexible spacers. A recent review describes much of this work in detail [7.41]. The effect of varying the length of flexible spacers has also been reported in [7.42], with measurements of both viscosity and alignment of a polymer of the form of ⟨66⟩.

$$\left[-(CH_2)_m - \overset{\overset{O}{\parallel}}{C} - O - \bigcirc - \overset{\overset{O}{\parallel}}{C} - O - \bigcirc - O - \overset{\overset{O}{\parallel}}{C} - \right]_n \qquad ⟨66⟩$$

When m was as large as 12 no orientation could be obtained, and the melt was very viscous. As m decreased, so did the viscosity, whilst the diamagnetic susceptibility increased. For $m = 5$ and 6, $\Delta\chi$ was comparable with typical small molecule values. The time taken to achieve full alignment, and the strength of the magnetic field necessary also increased with m. However, in some chiral compounds containing flexible spacers, with pitches comparable with the wavelength of light, a field as high as 16 T was insufficient to unwind the helix [7.43]. Nor did all the apparently nematic polymers examined in this same study

Fig. 7.14 (*a*) Development of orientation with time in a random
copolyester ⟨32⟩ of molecular weight 8600 at the different applied magnetic
fields indicated. (*b*) Orientation data for a range of molecular weights and
field strengths normalised with appropriate values of $\mathcal{S}_{max}$ and τ (see text).
[7.44] Symbols correspond to molecular weights as: $+\,4600$, ◆ 5000, △
8600, ■ 14400.

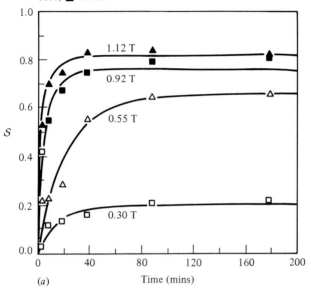

(*a*)

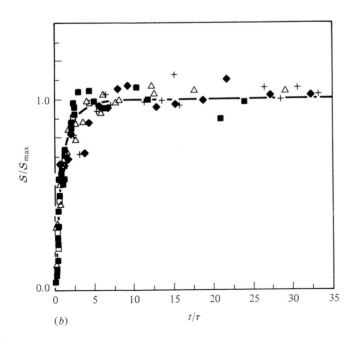

(*b*)

show alignment. It was suggested that this was because crystallization dominated over alignment.

Clearly, in order to understand the measured $\Delta\chi$, it is necessary to know the degree of order which the field is producing. Moore and Denn [7.41] have studied the relationship between the order parameter, S, (measured by X-ray diffraction) and the field strength for the random copolyesters $\langle 32, 37 \rangle$. Subsequently, Anwer and Windle [7.44] have

Fig. 7.15 (a) X-ray diffraction pattern showing the orientation achieved in a random copolyester of number average molecular weight 14 400 after 1 h in a magnetic field of 5.6 T. The field direction is vertical. (b) A fracture surface of a sample oriented in a field of 1.1 T for 1 h showing the high level of fibrillar alignment with the magnetic axis (vertical). (c) A fracture surface of a similar sample given the same thermal treatment but without the magnetic field. [Courtesy A. Anwer, unpublished work.]

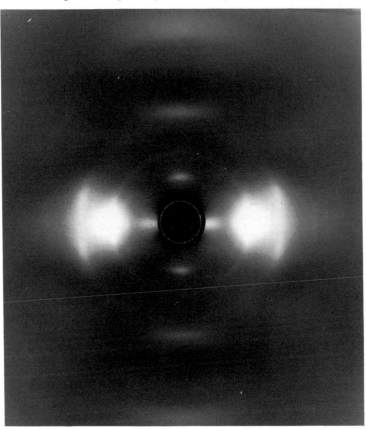

(a)

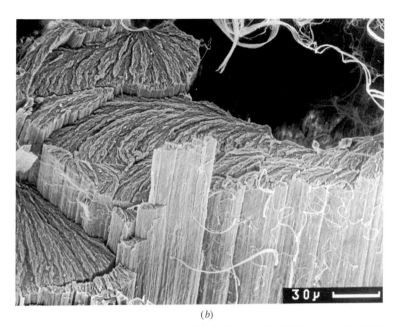

(b)

(c)

examined the effect of molecular weight as well for the random copolyester ⟨32⟩. The order develops rapidly at first with time and levels off towards a maximum value which depends on the field strength. The field strength also controls the rate of orientation. Figure 7.14(*a*) shows the development of orientational order with time, expressed in terms of the order parameter S, for a range of field strengths. It is apparent that the orientation–time curves can be fitted by the relation:

$$S = S_{max}[1 - \exp(-t/\tau)]$$

The value of S_{max} is the plateau level of orientations reached after long times. It depends on both the molecular weight and field strength. τ is a characteristic time for the process which describes the rate of orientation for a given set of conditions. Data for a range of molecular weights and field strengths normalized using the values of S_{max} and τ for each set of conditions are shown in Fig. 7.14(*b*).

Over a range of number average molecular weights from 4600 to 14400, and field strengths up to 1.1 T, the initial rate of orientation is proportional to the square of the field strength and to the reciprocal of the square of the molecular weight.

The maximum orientation obtained in fields up to 1.1 T was 0.8, but, at the higher field strength of 5.6 T, $S = 0.97$. The quality of orientation obtained is apparent from the transmission X-ray diffraction pattern Fig. 7.15(*a*), and the appearance of the transverse fracture surface in Fig. 7.15(*b*), which shows fibrillation, is compared with a similar surface from an unoriented specimen in Fig. 7.15(*c*).

It has also been found [7.41] that specimens which had been held in the field for short periods subsequently relaxed to a state with a far lower value of S when the field was removed from the molten polymer, than those held for longer times, even though the difference in initial order parameter (i.e. when relaxation started) was not that great.

7.5   Instabilities

Many types of instabilities have been identified for small-molecule liquid crystals, which may be associated with flow alone, flow combined with the presence of an electric field (electrohydrodynamic instabilities) or equivalently a magnetic field, or due to convection. Reviews of these various phenomena can be found in [7.45, 7.46]. Usually the instabilities are such that fluctuations in director orientation are stabilized, for example by the flow, to give rise to a periodic pattern – typically stripes, rolls or a square grid.

One of the most important parameters controlling the formation of these instabilities is the relative magnitudes of the various Leslie coefficients. It can be shown that a stable inclined alignment of the director under shear flow can be achieved only if α_2 and α_3 have the same sign. Similar, although more complex, conditions relating to the α's can be derived for other situations to determine stability/instability criteria involving flow. And because, for compounds such as PAA $\langle 2 \rangle$, the α coefficients are well established experimentally, the theoretical models can be tested readily for small-molecule materials.

For liquid crystal polymers the situation is much less well advanced. As we have seen the α coefficients tend not to be known. Furthermore, the uniform alignment necessary as a starting condition for the observation of the instabilities can rarely be achieved. Nevertheless, some electrohydrodynamic instabilities have been identified in both lyotropic and thermotropic systems.

In attempting to measure the Fredericks transition in PBLG solutions, Meyer and coworkers discovered the appearance of transient 'stripes' associated with the application of a magnetic field [7.47]. The stripes developed when the configuration for the twist-Fredericks transition (Section 7.2.3) was set up and a large magnetic field was suddenly switched on. Their analysis, expressed in terms of the Leslie coefficients, showed that the periodic director distortion underlying the stripe pattern has a faster response time than a uniform mode. The alternating stripes indicate regions with opposite senses of director rotation and associated with the consequent molecular rotation is a pattern of flow cells. The effective (shear) viscosity of the comparatively short wavelength stripes is now lower than the corresponding bulk (rotational) viscosity, favouring stripe formation.

The growth rate of any wavelength of stripe will be determined by the competition between the elastic distortions associated with the periodic director distortions, and the viscosity. For example, long wavelengths pay a low elastic energy penalty but have a high associated viscosity. It follows, therefore, that measurements on this stripe instability enable measurements of the (ratio of) elastic constants to be made (see also Section 6.2). It further transpires that, despite this particular instability not having previously been observed in small-molecule liquid crystals, it can be found in such a common and extensively studied material as MBBA $\langle 3 \rangle$. It had not been detected earlier because the transient response is very short lived.

Turning now to electrohydrodynamic instabilities in liquid crystalline polymers, some of the patterns well studied for small-molecule materials have been found in polymeric mesophases also. The work of Krigbaum and coworkers is notable in this respect. Working with a random copolyester ⟨33⟩ it was possible to create what are known as *Williams' domains* in an unoriented sample, either with AC fields up to 100 Hz or, more easily with DC fields (see e.g. [7.48]). The speed at which the domains formed increased with increasing temperature (i.e., decreasing viscosity). Some of the other instability patterns, such as the 'dynamic scattering mode', familiar to small-molecule liquid crystal experimenters, proved impossible to achieve because uniform initial orientation could not be attained by known techniques. The high field turbulence pattern was obtainable in the same copolyester, closely resembling that seen in nematic PAA, with a rise time of ~ 1 minute. As with the development of uniform orientation under electromagnetic fields, response times for these instabilities depend on the melt viscosity. This leads to the possibility of observing novel precursor structures, too short-lived to be observed in small-molecule liquid crystals.

Although the dynamic scattering mode has not yet been achieved in a main-chain polymer, it has in a siloxane side-chain polymer. In this case a low frequency (< 300 Hz) electric field of above a threshold strength $\sim 10^6$ V m^{-1} (but which was temperature dependent) produced the turbulent, light scattering pattern. Higher frequencies led to simple homeotropic alignment.

This section cannot provide an exhaustive coverage of all the instabilities observed to date. The appearance of instability textures seems very common under a wide variety of geometries, and care must always be exercised in deducing from their appearance the actual underlying molecular organization.

8 Practical aspects of liquid crystalline polymers

8.1 Application of liquid crystalline polymers – two perspectives

During the 1980s, research and development of liquid crystalline polymers (LCPs) has been one of the most rapidly growing areas of materials science. It represents an encounter between two fields; one being liquid crystal science which can be traced back to the work of Reinitzer in 1888, and the other polymer science which was probably not recognizable as such until the 1930s. However, the perception of what a liquid crystalline polymer *is* still depends to some extent on one's own background.

Where experience stems from small-molecule device materials, liquid crystalline polymers are most readily appreciated as conventional polymer backbones, siloxanes, methacrylates etc., onto which mesogenic groups have been grafted as side chains, or into which they have been incorporated as rigid sections within the otherwise flexible backbone. The polymeric character is seen as a means of achieving mobile and thus field-active mesogenic groups in the solid state. Furthermore, polymerization enables suppression of crystallinity, which means that the liquid crystal orientation can be frozen-in on cooling while preserving the optical texture as written.

On the other hand, from the traditional polymeric perspective, liquid crystalline polymers represent a new dimension of processibility whereby comparatively standard fabrication methods can lead to materials with a unique range of properties for general structural applications. The liquid crystallinity is associated with molecular chains which are uniformly stiffer than in conventional polymers. This stiffness is usually achieved by increasing the aromatic content of the molecules and connecting the groups so as to optimize chain straightness. Thus, whereas the device scientist has taken the small molecules with which he is familiar and dangled them on, or within, flexible polymer chains, the polymer scientist, with such chains as his natural starting point, has stiffened them up until they become liquid crystalline in their own right.

It is not surprising that fuller integration of these two approaches will open up new areas of application of LCPs. Devices are being built where the liquid crystallinity plays its main role at the fabrication stage, the activity being derived from a much wider range of electronic processes than the field orientation of the molecules *per se*. Correspondingly, there are several possible ways in which the processing of structural polymers will benefit from the application of electric and magnetic fields.

8.2   Liquid crystalline polymers as structural materials
8.2.1   *Stiffness and strength*

Of all the properties which are desirable in a high performance structural material, stiffness and strength are of central importance. In polymers, stiffness has often been a weak point compared with metals and other inorganic materials. Even highly drawn Nylon fibre has an axial modulus of only 5 GPa, whereas steel has a value of 210 GPa, in all directions, and boron fibres in the region of 350 GPa. However, it is important not to lose sight of the fact that diamond, with a $\langle 001 \rangle$ modulus of 1150 GPa, is in one sense the ultimate cross-linked organic polymer. The challenge is thus to persuade polymer chains to behave as if they were one-dimensional diamond. The degree of molecular organization necessary has now been achieved for a range of different polymer molecules, and the exploitation of liquid crystallinity has provided one of the significant routes to that success.

It is important to appreciate that chains, like pieces of string, will be much stiffer in tension than compression, and the account in the following sections will concentrate first on the achievement of high *tensile* stiffness and strength. There are several different ways of defining strength, particularly where a material undergoes considerable plasticity before final failure. However, in the case of high stiffness materials at least, it is taken as the stress necessary to produce rupture.

8.2.2   *Criteria for chain stiffness*

In considering the mechanical properties of individual molecules, it is helpful to distinguish between two contributions to the flexibility of the chain. There is firstly the contribution from the flexibility of individual backbone bonds in terms of both bond stretching and the distortion of the angles between different bonds at a given atom. For many molecules however, by far the greater contribution to flexibility stems from rotation about single backbone bonds which enables the chain to take up a wide range of trajectories.

Although our primary interest is in liquid crystalline polymers, it is instructive to consider the polyethylene molecule as a first example. If it is free to achieve thermal equilibrium, at a temperature above its melting point, the molecule will squirm into every conceivable shape, and the length between the ends will be changing between the limits of zero when the ends happen to touch, and the stretched out straight length of the molecule. An illustration of a polyethylene molecule in such a random conformation is shown in Fig. 8.1(*a*). If the ends are grasped in some way and eased apart against the randomizing influence of the thermal motion, then the molecule can be seen to have elasticity associated with a decrease of entropy on extension. The forces involved are not very high, and a material containing such random chains will have a low elastic modulus typical of a rubber. A typical value may be 10 MPa.

Rotations about backbone bonds lead invariably to changes in energy. There is frequently a single setting corresponding to the minimum energy, which in the polyethylene molecule is called the *trans* state. If the chain is built with all backbone rotations set at this minimum energy angle, then its overall trajectory will be straight (Fig. 8.1(*b*)). Furthermore, in the case of polyethylene, the lowest energy

Fig. 8.1 Molecular models of (*a*) polyethylene, random chain at 300 K, (*b*) polyethylene extended conformation, all bond rotations set at minimum energy angles, (*c*) polypropylene 3/1 helix, bond angles again at minimum energy rotations.

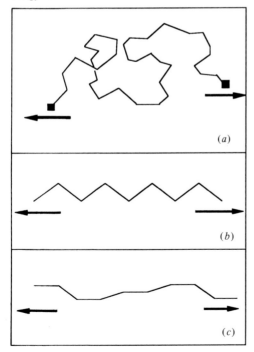

(*a*)

(*b*)

(*c*)

conformation also corresponds to the longest chain possible. This factor has important consequences for mechanical properties, as any further increase in chain length will have to be at the cost of distortion of the bond lengths and the tetrahedral bond angles, both processes demanding significant energy input. The force constants for these distortions, determined by a number of techniques such as spectroscopy, enable the axial modulus of the chain to be calculated. A recent value obtained by Klei and Stewart [8.1] is 320 GPa, a value perhaps 30000 times greater than that typical of an isolated random trajectory molecule in the rubbery state.

For polyethylene, the ultimate axial stiffness would appear to be achieved when the molecule is in its lowest energy conformation, so giving the longest possible chain with a straight trajectory. However, for other molecules, the lowest energy conformation may not necessarily be the longest. For polypropylene, the most stable conformation is a 3/1 helix with every alternate backbone bond rotated in the same sense by 120°. Figure 8.1(c) shows the polypropylene molecule in this conformation. The chain has the same number of monomer units as the polyethylene one in (b); but it is now shorter by around 13%. Axial stress will thus initially extend the chain mainly by the easier process of bond rotation, and the molecule will be less stiff than polyethylene. Sakurada and Kaji [8.2] have measured the axial modulus of helical polypropylene in a crystal to be 34 GPa, about 10% of the extended conformation value. One has only to consider the possibility of coiled-spring type helices to see that the average trajectory being straight is not by itself a sufficient criterion for maximum stiffness.

These simple molecules illustrate the important principle that for optimum axial stiffness, the molecules must not only be straight and aligned with the tensile axis, but also be in their most *extended* conformation with the backbone bond rotation angles set to give the longest chain possible.

8.2.3   *Making straight molecules*

Even in the case of a molecule such as polyethylene in which the lowest energy conformation is the most extended one, it is unrealistic to hope that the extended chain conformation can be achieved by dint of careful cooling. Even if cooling could be carefully prolonged, it would be interrupted by crystallization and subsequently the glass transition. However, whereas the formation of crystals effectively precludes further rearrangement of the chains, the crystallization process itself is an effective mechanism of self organization.

The molecules are locked in the extended-chain conformation, and held in perfect parallel alignment with each other. The difficulty is that the crystallization process is too localized and only comparatively minor sections of a high molecular weight chain are incorporated in any one crystal. The entanglements of the melt are thus not so much removed as concentrated into regions between the crystals where they give rise to low modulus, non-crystalline regions. The axial stiffness of the crystals can be high and they can be aligned very well by processes such as fibre drawing, however if they are separated by comparatively soft non-crystalline regions, the overall stiffness of the fibre will remain low. For conventional polymers, of which polyethylene remains the prime example, high axial stiffness requires the unravelling of the molecules either before or after crystallization. This is achieved by a variety of advanced mechanical processing methods, which may involve gels, flowing melts or solutions. The role of crystallization is to lock chains in their unravelled, high stiffness conformation. This topic is extensively reviewed in the literature by authors such as Capaccio *et al.* [8.3], Calundann *et al.* [8.4] and Lemstra *et al.* [8.5].

Thus, if one is to make axially stiff materials from flexible-chain polymers, it is necessary to both *organize* and *lock* the molecules in their extended-chain conformation.

The alternative approach is to design molecules which are intrinsically rigid and straight. This amounts to either a restriction of all rotational freedom or arranging that those bonds around which rotation is possible are as near as possible parallel to the chain axis. For example, the molecule of a ladder polymer shown in Fig. 8.2(*a*) is chemically straight with no freedom for conformational motion. In the case of poly(*p*-phenylene) (Fig. 8.2(*b*)), there is comparatively free rotation about the bonds connecting the phenylene groups, and yet this freedom does not lead to flexibility of the straight molecule. What little flexibility there is, will come from bond bending. Both these molecules are examples of *rigid chain* polymers and potentially have axial moduli greater than that of linear polyethylene, because each has a greater density of backbone bonds parallel to the chain axis. Poly(*p*-phenyleneterephthalamide) (PPTA), the molecule of Kevlar, has one bond of the amide link group at a substantial angle to the axis of the straight molecule (Fig. 8.2(*c*)). Rotational freedom about this bond is limited by virtue of electron resonance with the double bond of the neighbouring carbonyl group, which gives it some double bond character. However, even with the central bond set to give a linear link, different settings about the phenyl–carbon and phenyl–nitrogen bonds can lead to a side-stepping of the chain which, if not coordinated, will cause it to wander away from linearity. Two chains of the Kevlar type are compared in Fig. 8.2(*d*), one with all the phenyl–carbon bonds set at 0°, the other with their two equivalent settings of 0° and 180° populated at random. Again it is clear that optimum properties will require the all 0°, extended conformation which is stabilized in the crystal form of the polymer. The limited flexibility of this type of molecule has led to its classification as *semi-rigid chain*. Similar

Fig. 8.2 (*a*) polyacene ⟨67⟩. (*b*) Rotation of a *para* linked phenylene group. (*c*) Axial rotations in the repeat unit of poly(*p*-phenyleneterephthalamide) ⟨44⟩. (*d*) Comparison of an extended conformation in an aromatic polyamide or polyester, with a conformation which 'side-steps' up and down at random.

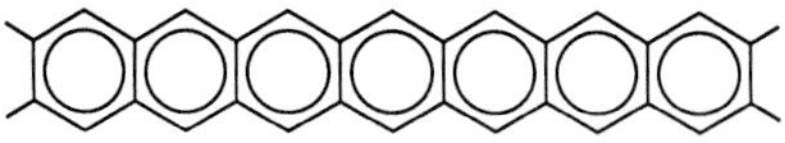

⟨67⟩

(*a*)

⟨15⟩

(*b*)

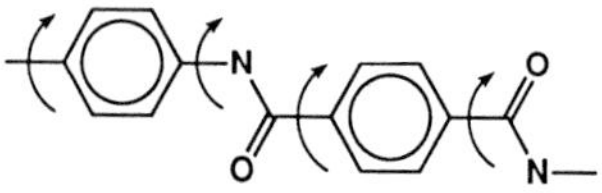

⟨44⟩

(*c*)

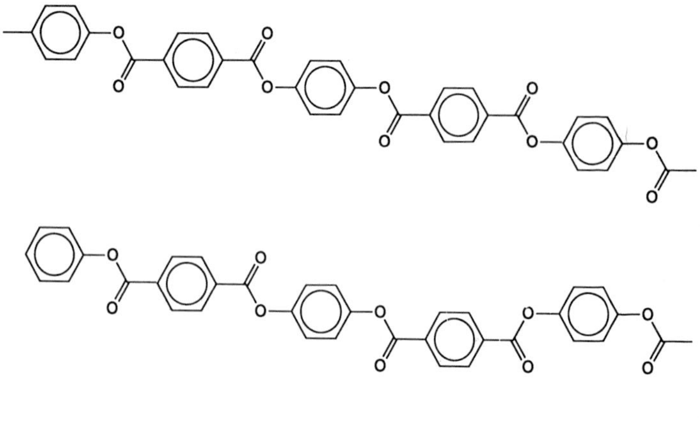

(*d*)

arguments apply to linkages with closely related geometry such as the ester one which has been discussed in detail in Chapter 3, especially Fig. 3.2.

Both rigid and semi-rigid chains tend to form liquid crystalline phases, and it is this liquid crystallinity which provides an efficient means of obtaining the high degree of molecular alignment necessary for a stiff material. Additionally, for the semi-rigid type, the required organization involves an element of conformational ordering to ensure the molecules are held in their axially stiffest and straightest state.

For rigid chains, the challenge is to achieve tractability so that the intrinsically stiff molecules can be aligned both with each other and with some external axis such as a fibre axis. As such molecules typically have very high glass transitions and melting points, the route to liquid crystallinity is lyotropic, i.e. through the addition of appropriate solvents. Once in the liquid crystalline state, the flow fields necessary to give a high degree of axial alignment are comparatively modest. For example, the fibres made from lyotropic solutions of poly(*p*-phenylenebenzobisthiazole) (PBTZ) ⟨45⟩ and poly(*p*-phenylene-benzobisoxazole) (PBO) ⟨46⟩ do not appear to be crystalline in the sense that there is no detectable longitudinal register between the chains. It seems that crystallization subsequent to solvent removal is not especially important to the achievement of good axial properties. However, in the case of semi-rigid chains both liquid crystallinity and subsequent crystallization appear to be important steps towards the required level of organization. Thus, the liquid crystalline phase ensures that the molecules are comparatively straight and aligned parallel with each other. It also retains the essential liquid-like mobility which enables the molecules to rotate readily and align to an externally applied field. The crystalline phase, in addition to locking in the liquid crystalline alignment, will reduce the frequency of local conformational defects, such as side steps, along the chain, and thus further optimize the axial stiffness. This second stage, although desirable, is not necessary, and there are examples of high modulus thermotropic fibres based on random copolymers, which show good axial properties despite very low levels of crystallinity. In this case the glass transition provides most of the essential locking component, while a proportion of frozen-in conformational defects are apparently tolerable.

8.2.4 *Axial stiffness of fixed conformation molecules*

The forces between atoms are well-enough understood to enable the axial stiffness of an isolated molecule to be calculated. Where a molecule is in its longest possible conformation so that rotations about bonds can only shorten the chain, then given the molecular geometry, it is comparatively straightforward to determine the stiffness in terms of the energies associated with distortions of bond length and bond angle. The method of *molecular mechanics* (see for example Burkert and Allinger [8.6]) has been developed to find the energies of the different molecular conformations, although its energy parameterization means that it is also able to predict mechanical properties. Energy parameters are listed as stretching and bending distortion constants, k_s and k_b, for a whole range of atom and bond types. Combined with similar relations for bond rotation energies, they are usually embodied in molecular modelling software. To a first approximation, the equations for stretching and bending distortion energies are of the type:

$$U_s = 0.5k_s(l-l_0)^2$$
$$U_b = 0.5k_b(\theta-\theta_0)^2$$

where l_0 and θ_0 represent the equilibrium values of bond length and bond angle, and l and θ the distorted values. A third component, the so-called stretch–bend term, is also important. It accounts for the influence of bond angle and bond lengths on the direct interaction energy between the atoms on either side of the one at which the bond angle is being considered and is described by a stretch–bend energy relation which incorporates its own constant, k_{sb}.

$$U_{sb} = k_{sb}[(l_1-l_0)+(l_2-l_0)](\theta-\theta_0)$$

Here, l_1 and l_2 are the distorted lengths of the two bonds making the angle θ.

The art of molecular mechanics is to determine the appropriate constants, a task known as parameterization, from as diverse a range of sources as possible. A given bond combination will be studied, principally using spectroscopic methods, in several different molecular environments, and the results combined to give the appropriate parameter. It should be noted that mechanical data themselves may also be used to refine the parameter tables, and that there is not, as yet, complete agreement as to which is the best procedure to use, although with the simpler chains at least the basic parameters give a satisfactory agreement with experiment. For example, currently available energy parameters for polyethylene drawn from work by Sorenson *et al.* [8.7] give a value for the axial modulus of a polyethylene crystal of 340 GPa.

Not surprisingly, calculations for different extended-conformation molecules yield different ultimate moduli. However, the most significant factor is the packing density of the chains in the polymer which is maximized by:

(a)   selecting molecules with non-backbone atoms (or groups) of minimum size,

and

(b)   choosing backbones of the highest possible unsaturation to reduce the number of substituents.

Figure 8.3 is a plot of theoretical axial moduli (measured values in the cases of diamond and graphite) against the effective density of polymer chains *taking into account only the backbone atoms*. This parameter is the density reduced in accord with the proportion of atomic mass not actually in the backbone or network. For example, the effective density of polyethylene is taken as the actual density $\times 12/14$

Fig. 8.3 Plot of calculated axial modulus for organic materials against the effective density which takes into account the mass of the backbone atoms only.

PBO	[8.8]	poly(benzobisoxazole)	⟨46⟩
PPTA	[8.8]	poly(*p*-phenyleneterephthalamide)	⟨44⟩
PE		polyethylene	
PDAs	[8.9, 8.10]	polydiacetylenes	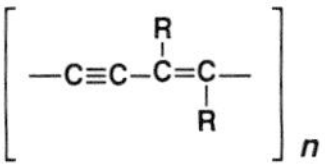 ⟨68⟩

$$\left[-C{\equiv}C{-}\underset{\underset{R}{|}}{\overset{\overset{R}{|}}{C}}{=}C{-} \right]_n$$

with, (in order of increasing chain density):

(*a*) [8.10]   R = —CH$_2$—N⟨

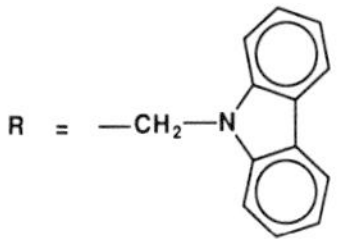

(*b*) [8.9]   R = —CH$_2$—O—C(=O)—N—⟨⟩

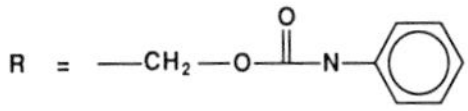

(*c*) [8.10]   R = —CH$_2$—O—C(=O)—N—C$_2$H$_5$

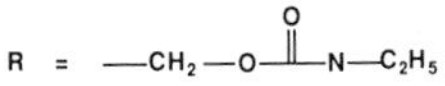

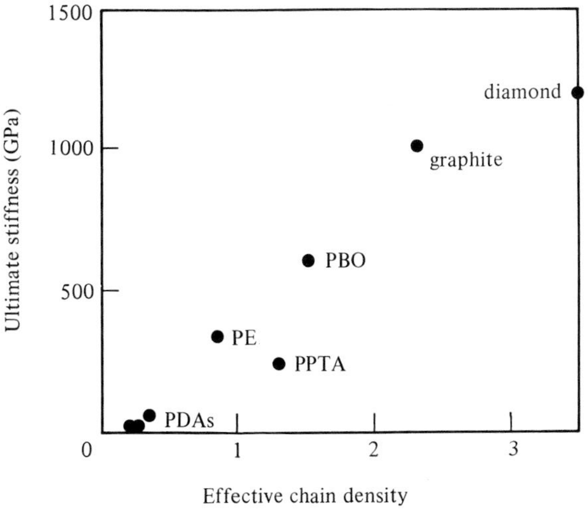

Table 8.1 *Values of axial tensile moduli* (*E*) *and specific moduli* (*modulus/specific gravity*, *E*/SG) *for a series of organic fibres*

Fibre	E(GPa)	E/SG
Kevlar 29 ⟨44⟩	60	42
Kevlar 49 ⟨44⟩	120	83
Kevlar 149 ⟨44⟩	185	128
Ultra drawn polyethylene (Spectra)	170	175
PBTZ ⟨45⟩	330	210
PBO ⟨46⟩	365	230
HBA/PET copolyester ⟨33⟩	23	16
HBA/HNA copolyester ⟨32⟩	65	46
Carbon fibre (high modulus)	390	215

which is the ratio of the molecular weight of the backbone atoms to that of the total molecular weight per chain repeat. The fact that the calculated values of ultimate stiffness for polyethylene, and for PBO and PBTZ obtained by Weirschke [8.8], as well as the measured values for diamond and graphite (in the basal plane), lie close to the 1:1 line demonstrates the underlying significance of the backbone packing density in determining ultimate stiffness. The data for the poly-diacetylenes (PDAs – ⟨68⟩) are experimental values for molecules which can be made by solid state polymerization. The chain extension and orientation is more or less perfect and in good agreement with theoretical predictions, but the low values of stiffness are the result of the large side groups attached to the molecules. Of the polydiacetylenes (*a*, *b*, *c*), the one with the lowest backbone density (*a*), has only 17 % of its molecular weight in the backbone, the other two (*b* and *c*) have 21 % and 26 % respectively. It is interesting that the calculated value for PPTA (Kevlar) lies well below that of polyethylene, and does not fit the relationship. This is despite the fact that $\frac{2}{3}$ of each PPTA chain is aromatic, and that they pack very nearly as densely as the polyethylene ones.

Actual limiting values for moduli for various types of organic fibre are listed in Table 8.1, both as absolute values and specific values. Of these, only PPTA (Kevlar) has been in large scale commercial production for some time. The data for poly(HBA/HNA), PBO and PBTZ are for preproduction samples. It should be noted that the low crystallinity of the random copolymer, coupled with the less than complete rigidity of the molecule is likely to lead to increasing degradation of the properties with increasing temperature. Such behaviour is discussed in the next section (Fig. 8.5).

8.2.5 Thermal motions

The discussion in the previous section is based not only on the premise that the chains are in an extended conformation, but that their local thermal motions are insignificant. Such an assumption is only secure at low temperatures. Where one is considering liquid crystalline materials in structural applications at ambient temperatures and above, thermal motion can sometimes greatly detract from good axial properties. There are four important types of motion:

(a)   statistical variations about the equilibrium values of bond length and bond dihedral angle which have an effect on the chain length

(b)   statistical fluctuations about the minimum energy settings of the bond rotational angles

(c)   rotational transitions between different minimum energy conformations

(d)   local motions of any type which do not affect the chain length.

Vibration about the minimum energy bond lengths and dihedral angles (a) can be seen as occurring within an energy minimum as shown in Fig. 8.4(*a*). The energy valley is not symmetrical because ultimately it is harder to squeeze atoms together than to pull them apart. At absolute zero, all bonds will lie at the minimum. However, with increasing temperature there will be a distribution of bond parameters, and thus the chain lengths will have a probability distribution about the minimum as shown in Fig. 8.4(*b*). The higher the temperature, the broader this distribution will be. The fact that the valley is asymmetric means that the mean length will move to higher values with increasing temperature, and thus the chain length will increase. The dotted plots on the figure illustrate the effect of applying a tensile stress along the chain. The minimum occurs at a greater chain length and the distribution is correspondingly skewed, and the mean end–end length increased.

The essential difference in the case of statistical variations about bond rotational angle, (*b*), is that, assuming the chain is in the maximum length conformation to start with, the rotational fluctuations can do nothing but shorten the chain. The corresponding energy and probability sketches in Fig. 8.4(*c*, *d*), show that the asymmetry is now in the opposite direction so that higher temperatures will lead to a shorter mean chain length, and that the effect of tensile stress will be to lengthen the chain by rendering less favourable those states which make the greatest contribution to shortening the chain. This contribution to elastic deformation is entropically based, just as rubber

Fig. 8.4 Schematic representation of the energy–length (*a*), (*c*) and (*e*) and probability–length relations (*b*), (*d*) and (*f*) for an extended chain with: thermal distributions of bond lengths and dihedral angles, (*a*) and (*b*); thermal distributions of bond rotation angles, (*c*) and (*d*); thermal distributions of bond rotation angles but with competing energy minima which are accessible in the solid state, (*e*) and (*f*). In each of the diagrams the dashed curves correspond to the displaced energy and probability distributions resulting from the application of a tensile stress along the chain axis.

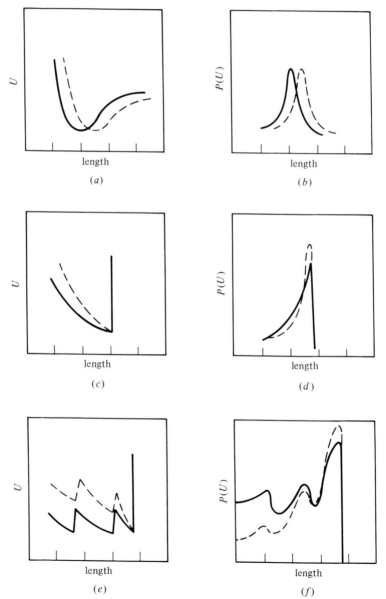

(*a*)

(*b*)

(*c*)

(*d*)

(*e*)

(*f*)

elasticity is, and will detract from the ultimate axial stiffness one might expect at absolute zero temperature and indeed predict by molecular mechanics. There is ample evidence that, in molecules such as aromatic polyesters, there can be thermal fluctuations between different, closely related conformational states within the solid state. As not all of these states will correspond to the longest possible chain, there is even greater opportunity for thermal shortening, and the relatively easy mechanical skewing of this distribution which will further reduce the axial stiffness. Fig. 8.4(*e, f*), is a representation of the energy–length and probability–length relationships for this situation.

It is well established that the axial coefficient of thermal expansion (CTE) for chain extended polymers is negative. It is the case for polyethylene crystals as well as for fibres of liquid crystalline polymers, and indeed also for directions within the basal plane of graphite crystals. These observations confirm that thermally induced rotational fluctuations which will shorten an extended chain are dominant over the extending effect of the bond length and angle fluctuations. Figure 8.5 shows measurements of the CTE and axial modulus made simultaneously on a well-oriented extrudate of a copolymer of hydroxybenzoic and hydroxynaphthoic acids (HBA/HNA) ⟨32⟩. It is apparent that the CTE is negative and that the axial modulus also decreases continually with increasing temperature. The glass transition of this material is at about 120 °C, and the rapid decrease in modulus

Fig. 8.5 Measurements of the axial tensile modulus (E) and coefficient of thermal expansion (CTE) obtained simultaneously from a drawn extrudate ($\mathscr{S} = 0.9$) of 73/27 HBA/HNA copolyester ⟨32⟩.

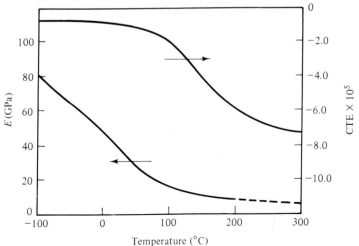

above this temperature is the consequence of much of the polymer being in the liquid crystalline state, the solid properties being maintained by a crystalline fraction of the order of 20%.

In virtually all chain molecules there are thermally induced fluctuations which do not in themselves affect the chain length. These may be rotations of side groups, or in the case of aromatic liquid crystalline materials, the rotation of backbone phenyl groups about their virtual bonds. Alternatively, rotation of aromatic groups such as the 2,6 naphthoic residue or rotation of a single bond adjacent to a phenyl group will lead to crankshaft motion of the chain backbone. These motions are illustrated in Fig. 2.1. One way of observing these motions is by using mechanical spectroscopy. It is possible to detect mechanical loss peaks under an oscillating tensile stress, which occur at different temperatures depending on the frequency of the applied stress. These peaks can be linked with the onset of specific thermally activated motions. Davies and Ward [8.11] have made dynamic mechanical measurements on benzoic–naphthoic random copolyesters and have identified three transitions, α, β and γ. A plot of the temperatures of the transitions as a function of strain rate (equivalent to frequency) is shown in Fig. 8.6. The α transition is the glass transition, and shows the very high activation energy characteristic of the process. The β transition is in the region of room temperature and can be identified with rotations of the naphthoic group and the associated changes to the

Fig. 8.6 Plot showing the positions of the mechanical induced transitions as a function of reciprocal absolute temperature and strain rate expressed as log frequency (f) for a 75/25 copolyester ⟨32⟩. (Redrawn from [8.11].)

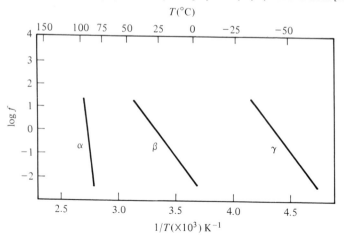

backbone conformation. The lowest temperature transition, designated γ, is identified with the rotation of the *para* linked phenyl groups in the backbone. Although these rotations are not necessarily associated with any change in the backbone conformations, their motion, because of the coupling between them and the flanking ester groups, will certainly be coupled with movement in the backbone. The data correlate quite satisfactorily with other measurements of molecular motion made on the same family of liquid crystalline polymers by Blundell and Buckingham [8.12] using dielectric spectroscopy and by Clements *et al.* with NMR [8.13]. It is significant however that the fibres or tapes used in the experiments may have less than perfect orientation, which, coupled with the fact that transverse properties are typically an order of magnitude lower than the axial ones, means that the observed behaviour will represent the thermal influence on the shear as well as the axial mechanical properties.

It is clear that in molecules where thermally induced rotations around backbone bonds, and in particular localized changes in conformation, lead to a shortening of the chain, there will be degradation of the axial stiffness with increasing temperature. Compare, for example, the onset of local motions represented by the data of Fig. 8.6 with the loss of axial modulus with temperature plotted in Fig. 8.5. Put simply, this thermal effect is the penalty which must be paid if semi-rigid random copolymer chains are used to achieve thermotropic liquid crystalline behaviour. Rigid chain molecules avoid this drawback, but they require, as a rule, lyotropic processing routes which are more cumbersome and less versatile.

8.2.6 *Tensile strength of fibres*

The tensile strength of any polymer with perfect axial alignment will depend on the strength of the individual molecules as well as the density to which they are packed. Estimates of the load at which a single covalent bond in a simple polymer chain would break are around 3×10^{-9} N (3 nN) [8.14], with a breaking strain of between 5 and 10 %. On this basis, the ultimate fracture stress might be expected to be related simply to the number of chains per unit cross sectional area. Such an approach assumes that a chain will break at its weakest point, and that similar breaks in nearby chains will coalesce to give final fracture of the fibre. For bond breaking loads of 3 nN, the theoretical tensile fracture stresses of the aromatic polyesters and polyamides, PBTZ, PBO, and polyethylene all come out to be about 20 GPa. That is assuming fibre samples in which the molecules are perfectly extended and aligned. There are however a number of reasons why the observed strengths of high strength polymer fibres are unlikely to approach 20 GPa.

In the first instance flaws, whether internal or on the surface, will act as stress concentrators and thus reduce the observed fracture strength. Their significance is readily apparent when measured values of fracture stress are seen to decrease as a function of increasing gauge length or diameter of the fibre specimen tested. However, in the case of fibres drawn from linear polymer molecules, such as those made using polymer liquid crystal routes where there is comparatively weak interchain bonding, flaws appear to have little detrimental effect. In such cases the transverse cracks are either effectively blunted by intermolecular shear or deflected to run along the direction of the fibre axis.

The strength of polymers in general depends on their molecular weight, and this relationship is particularly important when one is seeking to develop ultimate properties. The benefit of increasing molecular weight is shown for fibres of HBA/HNA random co-

Fig. 8.7 Plot of the tensile strength of fibres of HBA/HNA copolyester $\langle 32 \rangle$ against the reciprocal of molecular weight. M_0 is the average molecular weight of the repeat unit. (The data are redrawn from Calundann *et al.* [8.4].) ■ as-spun; ● heat treated.

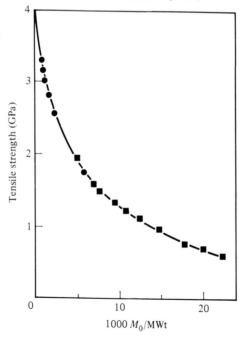

1000 M_0/MWt

polyesters ⟨32⟩ in Fig. 8.7. The tensile strength is plotted against the reciprocal of the molecular weight normalized to the mean molecular weight of the repeat units M_0. The higher molecular weight samples were obtained by annealing the fibres so as to continue the polymerization in the solid state. The data points lie on a continuous curve and show the underlying advantage of high molecular weight in obtaining optimum strength. This behaviour is to be contrasted with that seen in PPTA fibres in which annealing treatments tend to improve the axial orientation more than increase molecular weight, and thus enhance modulus rather than strength.

Another interesting aspect of the strength of these classes of fibre is the temperature dependence of the fracture stress. Figure 8.8 shows data for both PPTA (Kevlar) and heat treated HBA/PET copolyester fibres. A possible explanation of this behaviour is that the fracture of the highly stressed backbone bonds is promoted by thermal energy. If this is the case it is possible to write an Arrhenius type equation for the rate of bond fracture, $\mathscr{R}$, at a temperature, $T(\mathrm{K})$, as:

$$\mathscr{R} = v\exp\left[(-U+\beta\sigma)/kT\right]$$

The $\beta\sigma$ term represents the contribution of the applied stress to the reduction in the activation energy for bond fracture, β being the activation volume, and v is the frequency of thermal vibrations.

Fig. 8.8 The effect of temperature on the fracture strength of (i) a PPTA fibre ⟨44⟩ (Kevlar) [8.15]; and (ii) a heat treated random copolyester (60/40 HBA/PET) ⟨33⟩ [8.16].

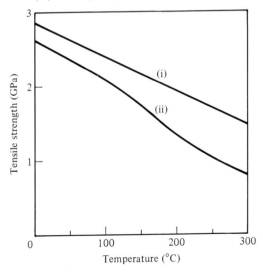

Termonia *et al.* [8.17, 8.18] and Termonia and Smith [8.19] have introduced a model for the strength of fully aligned aggregates of chains which incorporates the effect of temperature. They also treat the influence of the localized stress concentration around a single molecular break, and show how the enhanced rupture of other molecules in this region can lead to the growth of crack nuclei and eventual fracture. The specific computer model involves a parameter describing the shear strength parallel to the chains which controls the level of stress concentration associated with a break, while the influence of molecular weight, which determines the initial number of chain ends, is also explored. The model accounts for the observed decrease in strength both with decreasing molecular weight and with increasing temperature. It further demonstrates how creep elongation, prevalent in low molecular weight materials, can be the result of the cumulative effect of pockets of localized strain associated with the numerous chain ends. The fact that it is the *rate* of bond fracture which is determined by the thermally activated process, means that strength will also depend on time under load, or in the case of dynamic tests, on strain rate. For example, the model is able to describe the static fatigue results of Schaefgen *et al.* [8.15] for PPTA (Kevlar) which show for instance that the fibre will fracture after 100 days under a stress equal to 80% of its instantaneous (1 second) fracture stress.

8.2.7   *Compressive properties*

An isolated fibre is only really useful in tension. Compression leads directly to buckling, and any measured compressive properties will be derisory. Incorporation of fibres within a solid matrix to make a composite material opposes buckling and enables the actual compressive properties of the fibres to contribute to the properties of the bulk material. There is however a problem, in that even with the support of a matrix, the axial compressive strength of high tensile fibres is usually disappointingly low, as the polymer deforms by the mechanism of kink band formation. These bands are localized regions of the polymer in which the chains have cooperatively rotated away from the compression axis. It occurs when the shear strength across planes containing the molecules is low, and rotation enables the shear component of the applied stress to produce shear deformation, and indeed further rotation. The mechanism, shown in Fig. 8.9, can be seen when compressing a telephone directory along its pages. It will often 'fail' in compression by the kink mechanism. Composite matrices do not provide sufficient constraint to prevent the side step of the kink, and there is evidence for the lateral propagation of kink bands from one fibre to another. Figure 8.10 shows a series of kink bands in a fibre of PPTA which has been subjected to a compressive stress.

The measurement of compressive strength is not straightforward, and thus data from different sources must be compared with caution. However, Table 8.2 reproduces values compiled by Adams and Eby [8.20] for a wide range of fibres. In the case of fibres made by liquid

Fig. 8.9 Kink band mechanism for failure in compression.

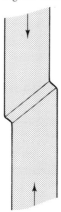

Fig. 8.10 Kink bands in a fibre of **PPTA** ⟨44⟩ produced by a compressive axial stress. They are approximately 45° to the fibre axis. Scanning electron micrograph, horizontal axis 15 μm (approx.). (Courtesy Dr D. J. Johnson.)

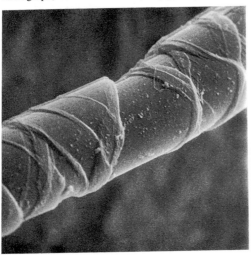

crystalline routes, the strength in compression is about an order of magnitude less than that in tension. Some carbon fibres show comparatively good compressive properties, and this would appear to be the result of the higher level of covalent bonding in the direction normal to the fibre axis. Increasing shear strength by bonding polymer molecules more strongly together, may well be a route to improved compressive behaviour, but initial work would suggest that occasional cross links between the aligned chains are ineffective.

8.2.8   *Quality of orientation*

With super-oriented materials, the price to pay for enhanced properties in the direction of the chain axis is a diminishment of the properties normal to the axis. Transverse weakness is most readily apparent as easy crack propagation along the axial direction, a phenomenon known as fibrillation. An extruded tube of liquid crystalline copolyester which has been cut with pliers is shown in Fig. 8.11. The extensive fibrillation is readily apparent. Not surprisingly, other mechanical properties are also degraded in the transverse direction; neither the modulus nor the coefficient of thermal expansion showing any advantage over an equivalent unoriented polymer. As well as causing a low value of tensile strength, the rather weak inter-molecular bonds mean that the shear properties on planes parallel to the molecular axis are also poor. Typical shear moduli for random copolyesters are in the range 0.5 to 1.0 GPa which is around 1 % of the axial tensile values. Where molecular alignment is less than perfect, an axial stress will lead to a small but finite shear stress parallel to the chain axes. However, because of the low shear modulus, this small

Table 8.2 *Tensile and compressive strengths of fibres* [8.20]

	Tensile strength (GPa)	Compressive strength (GPa)
Kevlar™ 49	2.6	0.48
Kevlar™ 149	2.5	0.45
Thermotropic polyester ⟨32⟩	2.6	0.20
PBO ⟨46⟩	5.8	0.40
PBTZ ⟨45⟩	4.2	0.40
Polyethylene (Spectra™ 1000)	3.0	0.07
Carbon (high strength)	3.2	2.88
Carbon (high modulus)	2.4	1.60
Carbon (ultra high modulus)	2.2	0.48

Fig. 8.11 Scanning electron micrograph of an extruded tube of a thermotropic copolyester ⟨32⟩, near to the point at which it has been cut across its axis. Horizontal dimension of photograph is 3 mm. (Courtesy C. M. Carr.)

Fig. 8.12 Correlation between measured values of longitudinal (E) and shear moduli (G) for fibres of thermotropic copolymers of HBA/HNA, ⟨32⟩. E_c is the longitudinal modulus determined by X-ray diffraction. The compositions are: △ = 73/27, +30/70 and ● 73/27 annealed. (Redrawn from [8.21].)

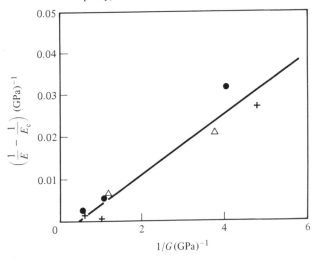

shear stress will lead to an appreciable contribution to axial strain and thus a significant reduction in stiffness. Troughton *et al.* [8.21] have shown that where the orientation is comparatively good, the approximation:

$$1/E - 1/E_c = (1/G)\langle\sin^2\Theta\rangle$$

is valid, where E is the measured axial modulus, E_c the axial modulus for the perfectly aligned case, G the shear modulus and $\langle\sin^2\Theta\rangle$ the mean value of the sines of the misorientation angles of the elements making up the distribution about the orientation axis. These authors applied this approach to highly drawn samples of thermotropic random copolyesters of the HBA/HNA system. They measured the values of E and G, which decrease with increasing temperature, using dynamic mechanical testing in tension and torsion respectively, while E_c was obtained from precise measurement of the position of the third meridional diffraction peak as a function of applied load at 120, 20 and $-80\,°C$. Their data are plotted in Fig. 8.12. If the equation above is fitted to these data, the slope of the graph gives 0.008 for $\langle\sin^2\Theta\rangle$ which indicates the high quality of the orientation of the samples. This is equivalent (again making an approximation justified because of the narrowness of the orientation distribution) to a macroscopic order

Fig. 8.13 Axial modulus as a function of the order parameter, S, for a series of fibre samples of a HBA/HNA random copolyester $\langle32\rangle$. (Redrawn from [8.22].)

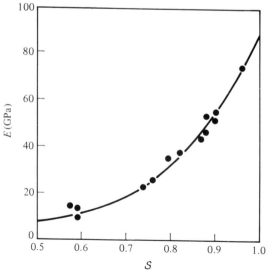

parameter (S or P_2) of around 0.99 which is the highest value ever reported for thermotropic liquid crystalline material. It is interesting to note that neither the composition nor the thermal history of the sample had much influence on the axial properties. Garg and Kenig [8.22] have measured axial moduli for the same type of polymer as a function of the extent to which an extrudate was drawn down on the line. The level of preferred orientation was obtained by wide angle X-ray diffraction, and their data for E are plotted against S in Fig. 8.13. It should be noted that the diffraction method for measurement of S becomes increasingly difficult to handle as the orientation approaches perfection. Nevertheless, the plot shows well the rapid reduction in modulus as the orientation decreases from the perfect value.

8.2.9 *Extrudates and mouldings*

The fact that thermotropic liquid crystalline polymers make extrusions and mouldings possible, means that one encounters more complicated orientation regimes than the simple axial alignment of a uniformly drawn fibre.

In extrudates the orientation in the skin layers is often much greater than in the core. This so called 'skin–core effect' is typical of liquid crystalline polymers but is lost if the extrudate is subsequently drawn down on line. Its origins have been described in Section 7.3.4. In brief the skin orientation is induced by the high shear gradients in the flowing polymer close to the die and the subsequent tensile load on this region just downstream of the die. The effect is clearly illustrated by measurements of Yamamoto [8.23] of the birefringence across the thickness of an extruded rod of a random copolyester of HBA/PET in the ratio of approximately 60/40. Figure 8.14(*a*) shows the effect of shear rate on the orientation profile. It is apparent that an increase in shear rate increases the quality of the orientation in the skin region without increasing the thickness of this region. It follows that the improvement in axial modulus of the extrudate which can be obtained by increasing the extrusion rate will be very limited. Figure 8.14(*b*) shows the effect of draw down on the orientation profiles. With a draw down ratio as low as 3.3 the skin–core distinction is obliterated, while higher ratios lead to a uniform improvement in orientation, and thus tensile modulus (Fig. 8.14(*c*)), which is in accord with the results of Garg and Kenig for HBA/HNA copolymers (Fig. 8.13). The increasing level of orientation at greater draw down means that the coefficient of thermal expansion gradually becomes negative as it tends towards the value characteristic of the perfectly oriented material. Yamamoto demonstrates this effect in a plot of thermal expansion coefficient against axial modulus for samples of different draw down ratio (Fig. 8.14(*d*)). An essentially zero coefficient of thermal expansion, coupled with high axial stiffness and strength mean that liquid crystal polymer extrudates are ideally suited to making protective jackets for glass fibres used for optical transmission. The coefficient of expansion of the polymer matches that of the silica glass and thus avoids optical signal

Fig 8.14 The effect of processing conditions on the orientation and properties of extrudates of a HBA/PET thermotropic copolyester ⟨33⟩. (*a*) Orientation profiles as a function of shear rate. (*b*) Orientation profiles for different draw down ratios. (*c*) The effect of draw down on axial stiffness. (*d*) The relationship between axial stiffness and thermal expansion. (From [8.23].)

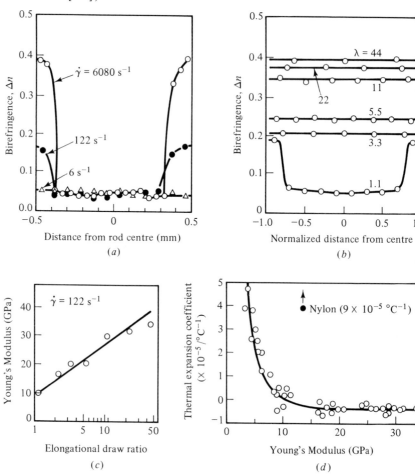

Fig. 8.15 The stiffness of liquid crystalline mouldings compared with conventional engineering plastics. LCP: liquid crystalline polymer. PES: poly(ethersulphone). PEEK: poly(etheretherketone). PBT: poly(butylterephthalate). PPS: poly(phenylene sulphide). (From [8.24].)

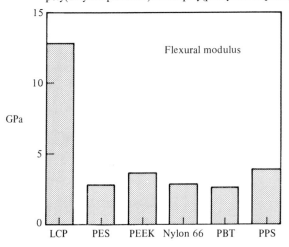

Fig. 8.16 The modulus of a moulding of thermotropic copolyester measured in the nominal flow direction for layers taken from different depths below the surface. The depth is expressed as the distance from the mid-plane of the moulding, t, as a fraction of the distance from the mid-plane to the surface t_0. The sample dimensions were approximately 200 mm × 50 mm × 3 mm thick. (Data from [8.22].)

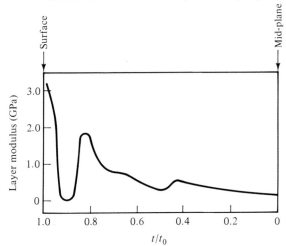

losses which can be caused by stresses introduced into the fibre as a result of differential thermal expansion.

In mouldings the flow patterns are generally much more complicated with both the orientation of the director and the order parameter changing with position. However, as with extrudates the surfaces of all liquid crystalline polymer mouldings are better oriented than the centres, and the fact that the material which has the highest modulus in the flow direction is concentrated within the skin regions means that it is especially effective in giving moulded bars high stiffness in bending (high flexural modulus). Figure 8.15 shows data reported by Cox [8.24] where the flexural modulus of a typical thermotropic copolyester is compared with values for a selection of non-liquid crystalline engineering thermoplastics.

Studies as a function of thickness have shown that the mechanical properties, like the structure, are exceedingly complex. Data from Garg and Kenig [8.22] reproduced in Fig. 8.16, show the modulus at different depths within a moulding of thermotropic copolyester. The high skin stiffness is apparent, and the two minima at approximately 0.9 and 0.45 of the semi-thickness appear to correlate with dark bands which can be seen on a polished cross section. These bands are possibly associated with the confluence of different 'streams' of the flowing liquid crystalline melt.

8.2.10   *Structural applications of liquid crystalline polymers*

Liquid crystallinity provides an important new route to polymer processing. Mesogenic fluids show highly efficient molecular orientation in comparatively modest flow fields, and thus provide a means of realizing the excellent intrinsic tensile properties of a straight polymer molecule. However, as various advantages are realized, there are often associated weaknesses to contend with, and in the final analysis liquid crystalline polymers will make an impact on the market not by showing any one outstanding attribute, but because of a particular combination of properties which make them the best material for a range of applications.

Fibres, whether made by lyotropic or thermotropic routes, show good axial strength and stiffness. The natural extension of the molecules in the mesophase is maintained in the solid, and lack of 100% crystallinity is not associated with loss of orientation. They find application, not only as reinforcing fibres in composites, but in heavy duty ropes and belting, and also in bullet proof clothing. Throughout the 1980s Du Pont's Kevlar has been the unchallenged commercial leader, although thermotropic fibres, which are simpler to fabricate, are beginning to make inroads on the market. The tensile properties, especially when expressed as specific values (i.e., divided by specific gravity), look impressive. Figure 8.17 highlights data for fibres of Kevlar, PBO and PBTZ, in comparison with other organic fibres, steel and glass. While the general trend shows high modulus associated with high strength, the stiffer forms of carbon fibre obtained by high temperature treatment tend to be rather weak. Carbon fibre composites do not give the radar transparency sought in some military aviation developments; also if they are burnt or damaged a fine suspension of particulate carbon can play havoc with electronic components. On the other hand, some grades of carbon fibre show very respectable compressive properties, an attribute which it will be difficult to achieve with a liquid crystalline polymer fibre.

The advantages of thermotropic polymers, usually random co-polyesters, start with their excellent mouldability. The viscosity of mesogenic melts is lower than that of conventional polymers of comparable molecular weight, and this means that moulds fill easily, the need for multiple gating is reduced, and cycle times can be

Fig. 8.17 Comparative plot of specific stiffness against specific strength for high performance organic fibres and other materials.

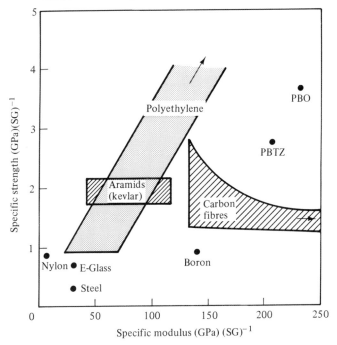

minimized. On the debit side is the problem that weld lines, produced where two flow fronts meet, detract from properties rather severely. They have to be carefully sited in non-critical positions if their effects are to be minimized. The low viscosity of liquid crystalline polymers means that they can accept high filler loadings, giving specific advantages with regard to both properties and cost.

The good mechanical properties of thermotropic mouldings at room temperature decay with increasing temperature and are effectively lost at the glass transition temperature T_g, as their random copolymer nature limits the crystallinity to too low a level to contribute significantly to useful mechanical properties. Furthermore, there is a steady loss of properties with increasing temperature between ambient and T_g, and this effect is particularly marked where dynamic transitions, such as the β transition in HBA/HNA copolymers, occur in this range. As discussed in Chapter 3, T_g can be enhanced by chemical modifications particularly those which both stiffen and kink the chains, but the fact that they cut back the crystallinity almost completely means that advantages given by the reduced processing temperature will be offset by the fact that mechanical failure will be total at T_g. New generations of mouldable liquid crystalline polymers can be expected to have glass transitions in excess of 200 °C, compared with the 120 to 150 °C of the materials which became available in the 1980s. There is also the possibility that high crystallinity thermotropic homopolymers could lead to materials where the ordered regions make a useful addition to the mechanical properties above the glass transition. One such polymer is the polyester based on terephthalic acid and chloro-hydroquinone ⟨28⟩. The possible 2,3 isomerism of the chlorine substitution means that it is not strictly a homopolymer, although the backbone repeat is completely regular. If the elevated temperature properties of these polymers are currently a little disappointing, they do show outstanding shape stability. This attribute is in part associated with the low coefficient of thermal expansion, but the limited level of crystallinity coupled with the fact that the crystal density is unlikely to be much greater than that of the melt is also significant. Mouldings thus reproduce the mould dimensions very faithfully showing also good resistance to warpage on cooling. Liquid crystalline polymers are thus ideally suited to applications where high precision is of the essence.

Polymers based on rigid molecules are reluctant to dissolve. This is shown by the fact that the solvents necessary for making lyotropic dopes are frequently vigorous, if not vicious. The low entropy of melting characteristic of a stiff chain polymer, is matched by a low partial molar entropy of dissolution. Liquid crystalline polymers thus show good resistance to solvents other than those which are able to form strong molecular associations with the chains. General environmental resistance is also associated with minimal water uptake which limits degradation in contact with molten solder, a great advantage in the construction of electrical components. Another aspect of the reluctance of these polymers to take up small molecules is their excellence as barrier materials, and there are many potential applications in the high performance packaging field.

While there are several exciting examples of what can be achieved with liquid crystalline polymers, ultra stiff and strong fibres, magnetically written materials etc., their assurance of a foothold in the plastics market place, stems more from a unique combination of somewhat more mundane properties. As an example of this, the electrical connector illustrated in Fig. 8.18, is made from a liquid crystalline polymer primarily because the material can be moulded accurately, withstands solder temperatures, and has adequate stiffness and strength. Fire and solvent resistance are additional advantages.

Fig. 8.18 Precision electrical connectors made from a thermotropic copolyester (Vectra). (Courtesy Hoechst Celanese.)

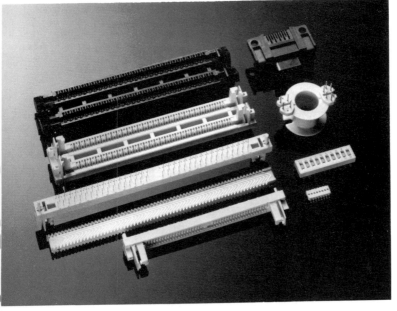

8.3   Liquid crystalline polymers as opto-electronic materials

8.3.1   *Electro-activity in organics*

The delocalization of electrons which can occur in some organic molecules, especially those with aromatic rings or other unsaturated bonds, means that the electronic distribution can be significantly influenced by various types of applied field, whether this be external, as with an electric field, or internal by virtue of the attachment of electron donating or withdrawing groups to the molecule. This means that while the synthesis of opaque, conducting polymers is possible, there is a wide range of effects which occur in molecules in which the electronic delocalization is of more modest extent.

The fact that either permanent or induced molecular dipoles interact with an applied field to orient the molecule is particularly useful. For weakly coupled molecules in a liquid, the effect is opposed by the randomizing influence of thermal motion, and although the orientation achievable by applying electric fields to polar liquids, as in the Kerr effect, may amount to a significant effect in the bulk, the actual orientation induced in each molecule is exceedingly small. The same applies to the Cotton–Mouton effect, the equivalent magnetic phenomenon. In a liquid crystalline phase, the orientational organization of molecules reduces the randomizing influence of thermal energy, the effectiveness of an applied field is thus much more marked, and the way is opened to building highly effective opto-electronic devices such as liquid crystal displays.

It should be noted that in some instances, liquid crystals are used in display devices not so much because of their ready field-orientability, but by virtue of their capacity for forming fine textures which scatter light strongly. It has proved possible to control the microstructure and hence its scattering power through local laser-addressed heating, and thus to create a writable device.

On another front, the delocalization of electrons can lead to non-linear interactions with light, in other words the response of a material to light, in terms of its refractive index for example, will depend on the intensity of the light. This property, which requires limited electronic mobility so that the material remains a transparent insulator, opens the way to optical–optical amplification and thus to logic devices.

8.3.2   *Usefulness of polymerization*

While the majority of liquid crystals used in devices are not polymeric, and electro-active polymers are not necessarily liquid crystalline, the 'overlap' region of liquid crystalline polymers opens up several intriguing possibilities for electro-optical materials.

(1) *Control of the fluid state.* The application of liquid crystals in devices depends largely on the field-orientability of the molecules. The intrinsic mobility of a liquid phase is important in this respect, and yet it leads to complications. The fluid needs to be contained in a cell, any charged species may tend to migrate towards the electrodes, and additions of other types of molecule, as in guest–host devices, are limited by their solubility in the liquid crystalline phase, with the additional possibility of subsequent phase segregation under an applied field. The combination of mesogenic groups with flexible polymer backbones, can do much to overcome these difficulties while maintaining the intrinsic orientability of the active groups. Such polymers consist either of mesogenic groups positioned at intervals along an otherwise flexible backbone, or as mesogenic side groups attached to a non-liquid crystalline backbone through flexible spacers. In the device field, the side-chain arrangement appears to be by far the most useful. The viscosity of the system is markedly increased by the presence of the backbone, and quite apart from the benefits in terms of reducing problems of cell leakage, it can be increased to the point where the material does not need cell containment at all. Furthermore, the introduction of a fraction of chemical cross links between the chains can produce an elastomeric liquid crystal with infinite viscosity and the potential to couple mechanical fields into the range of physical phenomena shown by conventional liquid crystalline devices.

The possibility of attaching other groups to the chain as well as the mesogenic ones, to form a copolymer in fact, opens the way to incorporating different types of active species into the material without restriction due to solubility limits, or risk of segregation. This advantage can be exploited to create multifunctional materials, where, for example, units might be introduced not only to induce a mesophase, but also to provide elements of non-linear, piezoelectric, dichroic or fluorescent behaviour. Copolymerization provides a means of controlling mesophase stability, both with regard to the isotropic transition and the intermediate ones such as smectic-to-nematic. Typically, control is exercised by reducing the number of mesogenic side groups per unit of backbone, or by mixing side groups which differ only with respect to the lengths of their flexible spacers.

(2) *Fixability of liquid crystalline microstructures.* When a small-molecule liquid crystalline material crystallizes there is normally a drastic change in microstructure. The formation of the crystals creates completely new structures which are often spherulitic as a consequence of a nucleation and growth process, and scatter light, making the solid opaque. With small molecules it is in fact exceptionally difficult to

preserve into the solid state, any structure which is characteristic of the liquid crystalline phase. There have been some reports that small-molecule mesogens can be quenched to form a glass, but it is difficult to exploit such a phenomenon on a routine basis, especially as the glass would normally have to be held at sub-ambient temperatures to prevent subsequent crystallization.

In side-chain liquid crystalline polymers, crystallization is generally inhibited so that cooling the mesophase leads to the preservation of its microstructure as a glass. Furthermore, the presence of the polymer backbone increases the glass transition temperature, and it is possible, through the selection of a sufficiently rigid backbone, to raise T_g enough so that the glass is stable for indefinite periods at ambient temperatures. An example of a backbone which provides such control is that based on poly(methylmethacrylate) (PMMA), where the mesogenic side chains in effect replace the methyl group attached to the ester. The T_g of PMMA is 105 °C, although full substitution with a mesogenic group such as: $[—(CH_2)_6—O—biphenyl]$ (cf. $\langle 77 \rangle$ in Table 8.4) will lower this by 50° or so. Partial substitution means that the T_g of the liquid crystalline polymer will be closer to that characteristic of the backbone.

The ability to fix a liquid crystalline texture into the solid state has many potential applications. Textures can be laser written and preserved, while processing the mesophase using force, electric or magnetic fields, can induce molecular orientation and optimize physical properties, such as non-linear optical (NLO) activity, in the glass. In this context we can see liquid crystallinity as an enabling phase which provides the means of generating highly specific structures in the solid state.

Although crystallization of chains with large mesogenic side groups is easily inhibited, the same is not true with main-chain mesogenic materials. However, it is possible to prevent, or at least considerably reduce, crystallization by making a random copolymer, an approach which has already been discussed in terms of lowering the melting point so that polymers can be processed without degradation. The very much lower crystallinity in such systems, coupled with the fact that the crystallites themselves are small and discrete, means that the glass preserves the main microstructural features of the mesophase, while light scattering due to those crystallites which are present is minimal, at least in thin film material.

(3) *Enhancement of molecular properties.* Conjugation between neighbouring aromatic groups along a polymer chain, can lead to continuous tracks along which the electrons are mobile. This is particularly

noteworthy in that it provides an important step in the design of conducting and semiconducting polymers [8.25]. However, liquid crystallinity is not as yet a normal route for the manufacture of these frequently intractable polymers. The non-linear optical (NLO) properties of sequences of conjugated aromatic rings, particularly with respect to what is known as the third order effect, appear to be quite dependent on molecular length and this can be enhanced by building appropriately conjugated main-chain polymers. Many NLO materials can be processed as mesophases and they now form a significant class of liquid crystalline polymers [8.26].

8.4   Non-linear optical (NLO) applications

Organics have distinct advantages as NLO materials. The electronic activity is especially fast, in the femto (10^{-15}) second range. It is also essentially lossless, which is not the case with either inorganic materials or multiple quantum well semiconductors such as those based on GaAs. The NLO properties of organic materials have been studied for a wide range of configurations such as: single crystals, liquid crystals, guest–host materials, polymers and liquid crystalline polymers.

The relevance of liquid crystalline polymers is threefold. Firstly, the molecular design requirements which will optimize NLO parameters, are just those which will lead to straight, rigid molecules, and thus liquid crystallinity. Secondly, a mesophase is an ideal medium in which the molecules can be aligned and poled (implying here the field induced alignment of permanent dipoles so that they lie parallel rather than anti-parallel). Thirdly, it is relatively easy to quench a polymer to a clear glass which is tough and can be formed into a film.

8.4.1   *NLO parameters*

The response of a material to an applied electric or optical field (**E**), is described by its polarizability (**P**). The magnitude of the polarizability for a given field is determined by various susceptibilities (χ) which are coefficients of the series:

$$P_i = \chi^{(1)}_{ij} E_j + \chi^{(2)}_{ijk} E_j E_k + \chi^{(3)}_{ijkl} E_j E_k E_l + \dots$$

where **P** and **E** are vector, and χ tensor quantities.

The coefficients, $\chi^{(2)}_{ijk}$ and $\chi^{(3)}_{ijkl}$, are the non-linear parameters in which there is so much interest. Where they are significant the polarizability of the material will depend on the intensity of the light. In this case the terms contributing to the refractive index are of the form $\chi^{(2)}_{ijk} E_k$ and $\chi^{(3)}_{ijkl} E_k E_l$, and are thus proportional to the square root of the light intensity and to the intensity itself respectively. However, it should be emphasized that the equation is also true for applied electric fields, so that where non-linear behaviour is present these fields will also influence the refractive index. The fact that an applied field influences the optical properties of the material opens up a wide range of potential device applications.

It is possible to write an equivalent relation for the individual molecular contributions to the bulk susceptibility. For a given active molecular unit, we thus have a molecular susceptibility of:

$$\mu_i = \alpha_{ij} E_j + \beta_{ijk} E_j E_k + \gamma_{ijkl} E_j E_k E_l + \dots$$

The relationship between the molecular β and γ terms and the macroscopic non-linear susceptibilities, $\chi^{(2)}$ and $\chi^{(3)}$, depends on factors such as the number of molecules per unit volume and local field corrections. However, the most significant parameter in linking the molecular and macroscopic effects is that of the preferred orientation. As the above equations are tensorial relations, the bulk susceptibilities will be dependent on the degree of alignment of the active groups. In fact $\chi^{(2)}$ will depend on $\beta \langle \cos^3 \varphi \rangle$, and $\chi^{(3)}$ on $\gamma \langle \cos^4 \varphi \rangle$, where φ is the angle between the principal axis of maximum molecular susceptibility and the equivalent axis for the bulk parameter. In a uniaxial material, the principal axis of the bulk material will be the molecular director.

It follows that the achievement of high NLO activity will depend not only on high values of β and γ at the molecular level, but on a high degree of alignment of the active molecular units. For second order effects, the preferred orientation must be *dipolar* in the sense that the longitudinal asymmetries of the aligned molecules must not cancel themselves out, in other words they must all point the same way, with orientations of $\varphi = 0°$ ($\cos^3 \varphi = 1$) not being cancelled by those of 180° ($\cos^3 \varphi = -1$). In the case of third order effects, where $\cos^4(0°) = \cos^4(180°) = 1$ then *quadrupolar* alignment is sufficient and the molecules do not need to be oriented with regard to sense. The important conclusion is that for materials to show significant second order NLO properties they cannot possess a centre of symmetry, in fact in crystallographic terms they must belong to a polar class. A nematic liquid crystalline phase cannot provide the necessary self organization to give second order properties. The phase needs to be *poled*, typically by an applied electric field, to break the nematic symmetry and orient all the molecular dipoles in the same sense. For third order effects, NLO materials do not require poling, and the necessary quadrupolar alignment can, at least in principle, be achieved by mechanical, surface or magnetic as well as electric fields. Symmetry considerations also show that NLO active materials will show some third order effects without any orientation at all, at about one-fifth of the value for the fully aligned state, whereas $\chi^{(2)}$ for randomly oriented material must necessarily be zero. These conditions are summarized in Fig. 8.19.

8.4.2 *Molecular design for high NLO activity*

The optimization of NLO behaviour in organics depends on two main factors: the design of the active molecular elements to maximize the β and γ parameters, and the organization of these elements within the structure so that their activity is transmitted in the most efficient manner to the material itself. Attention will be focused first on the design of the active molecular elements.

(1) *Second order effects*. A systematic investigation of the influence of molecular design parameters on the molecular second order coefficient, β, has been published by the group from Hoechst Celanese Corporation (e.g. Stamatoff *et al.* [8.27] and DeMartino *et al.* [8.28]). They developed a convenient method of assessing the value of β, based on the measurement of the shift of the visible/UV absorption on dissolution in different solvents, and applied it to a range of specially synthesized molecules. Some of their results are summarized in Table 8.3.

The basic molecular characteristics for the optimization of second order NLO are:

(a) large fixed dipole, which is achieved through the substitution of an electron donating group on one end of the molecule and an electron withdrawing group on the other. A molecule which is centrosymmetric will have a zero value of β and thus cannot show second order activity.

(b) freedom of electron movement along the length of the molecule. This property requires delocalized electron states,

Fig. 8.19 Influence of different orientation regimes on NLO parameters.

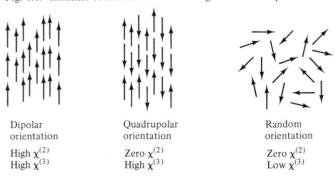

Dipolar orientation	Quadrupolar orientation	Random orientation
High $\chi^{(2)}$	Zero $\chi^{(2)}$	Zero $\chi^{(2)}$
High $\chi^{(3)}$	High $\chi^{(3)}$	Low $\chi^{(3)}$

Table 8.3 *Molecular structure control of β* (From [8.27] and [8.28])

	Structure	β at 1.9 μm $\times 10^{-30}$ esu
<69>	 NH_2—⬡—NO_2	5.7
<70>	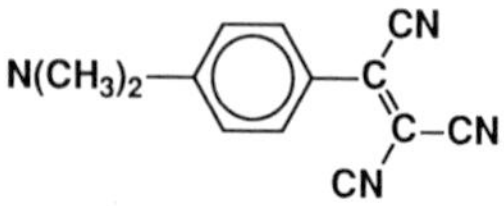 $N(CH_3)_2$—⬡—C(CN)=C(CN)—CN	21.4
<71>	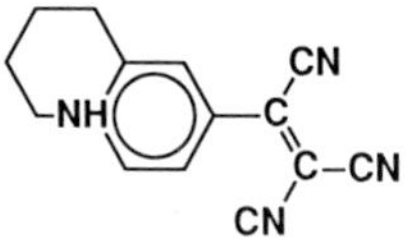 NH—⬡—C(CN)=C(CN)—CN	41.8
<72>	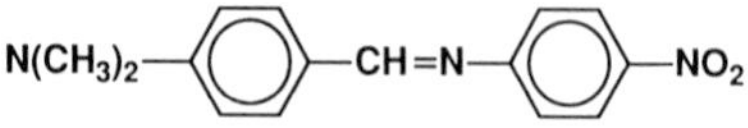 $N(CH_3)_2$—⬡—CH=N—⬡—NO_2	23.4
<73>	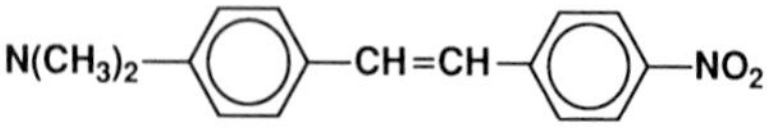 $N(CH_3)_2$—⬡—CH=CH—⬡—NO_2	60.0
<74>	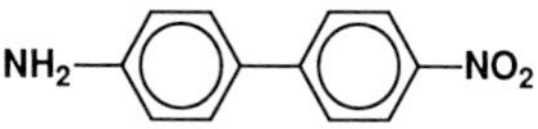 NH_2—⬡—⬡—NO_2	20.1
<75>	 NH_2—⬡—⬡—⬡—NO_2	50.7
<76>	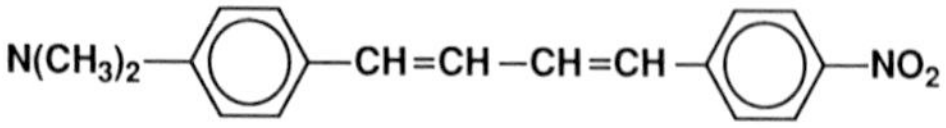 $N(CH_3)_2$—⬡—CH=CH—CH=CH—⬡—NO_2	111.2

such as the aromatic π states, which are conjugated along the molecule's length. The quality of the conjugation is dependent on both the persistence of the electron delocalization through the linking groups, and the mutual planarity of the rings.

(c)   long, straight molecule.

The data of Table 8.3 illustrate the importance of some of these factors. All the molecules have an electron donating group on the left and a withdrawing group on the right. The examples chosen illustrate the influence of three factors, the effectiveness of the end groups in imparting directionality to the induced dipole moment, the influence of planarity and the molecular length. Molecules ⟨69⟩ and ⟨70⟩ show the result of improving the electron donation and withdrawal efficiency. The cyano group is somewhat more effective than the nitro in withdrawing electrons. At the other end the dimethyl substitution on the nitrogen makes the group more nearly coplanar with the rest of the molecule increasing its donating efficiency, while its incorporation in a second ring in molecule ⟨71⟩ forces complete planar alignment. The azomethine linkage in molecule ⟨72⟩ twists the two rings with respect to each other, while the stilbene linkage of molecule ⟨73⟩ ensures planarity, effectively doubling β. The final three molecules illustrate the advantages of increasing the molecular length. It is apparent that this effect is greater than pro rata, which is important in that $\chi^{(2)}$ also depends on the number of molecules or active groups per unit volume, so that any doubling of length which produced only a doubling of β would be of no advantage. It is interesting to note that the factors which enhance the second order non-linear optical activity of the molecule are essentially those factors which encourage mesogenicity. Increasing axial ratio, planarity and charged end groups all encourage liquid crystallinity in small-molecule materials.

(2) *Third order effects.* Although the molecular criteria to optimize γ

Fig. 8.20 Calculations of γ as a function of molecular length of *trans* acetylene ($C_n H_{n+2}$). (From Prasad [8.30].)

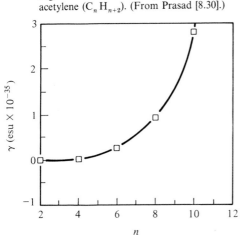

and thus $\chi^{(3)}$ have yet to be studied experimentally as systematically as for second order effects, a number of factors are emerging. In essence criteria for high γ are the same as those for high β, except that there is no need to generate an asymmetric electron environment through the apposition of donating and withdrawing groups. However, recent calculations by Medrano [8.29] would indicate that the presence of either donating or withdrawing terminal groups enhances the molecular susceptibility, γ, and the higher their efficiency, the greater the improvement. Calculations by Prasad [8.30] for a series of short-chain *trans* acetylene molecules (C_nH_{n+2}) show a very steep dependence of γ on length (Fig. 8.20). Although not apparent from the plot, the value of γ is marginally negative for C_2H_4. It is not surprising that given the criteria of length, straightness, and conjugation, attention has turned to the rigid-chain molecules of the type developed to give ultimate tensile mechanical properties. Of these PBO $\langle 46 \rangle$ and PBTZ $\langle 45 \rangle$ have received particular scrutiny under a programme initiated by the US Air Force (Ulrich [8.31]).

8.4.3   *NLO liquid crystalline polymers*

To date there are three distinguishable mesophase routes to the fabrication of NLO polymers: side-chain thermotropic, main-chain lyotropic and guest–host. In each case the liquid crystalline phase provides the unique combination of internal orientational order and molecular mobility, and is the ideal medium for field alignment, and in the case of $\chi^{(2)}$ materials, for poling.

(1) *Side-chain LCPs for NLO applications.* The fact that second order materials need to be poled as well as aligned, means that the molecular length cannot be excessive or the field response times are likely to be rather long, even in lyotropic solution. Hence long, straight polymer molecules with the appropriate terminations to give an asymmetric electronic environment are not the best route to maximizing $\chi^{(2)}$, and most effort is being focused on the development of side-chain materials.

The first question to be decided is whether it is possible to polymerize a molecular sequence, effectively adding the mesogenic side groups onto a backbone, without degrading a promising β. One particular comparison has been made by Stamatoff *et al.* [8.27], who measured β solvatochromically for the methacrylate monomer of 4-(12-do-decyloxy)-4'-nitrobiphenyl $\langle 77 \rangle$ and the resultant side-chain polymer. The solvatochromic method involves measurement of the frequency dependence of light absorption of the molecules in solvents of different and known polarity. The frequency shift with solvent polarity has been shown to be related to the same molecular factors as the second order

Table 8.4 *Comparison of β's measured for a monomer and the resultant side-chain polymer*

$$
\begin{array}{l}
\text{CH}_3 \\
\text{CH}_2=\text{C} \\
\text{O}=\text{C} \\
\text{O}-(\text{CH}_2)_{12}-\text{O}-\!\!\bigcirc\!\!-\!\!\bigcirc\!\!-\text{NO}_2
\end{array}
\qquad \langle 77 \rangle
$$

	Monomer ($\langle 77 \rangle$, with $n = 12$)	Polymer (Fig. 8.21(a) $\langle 78 \rangle$, with $n = 12$)
$\beta \times 10^{-30}$ esu	11	8

non-linearity for which it can provide a reasonable and easily obtained estimate. Their results are shown in Table 8.4. These values are considerably lower than were obtained for p-diphenylnitroamine shown as molecule $\langle 74 \rangle$ in Table 8.3. The oxygen bridge connecting the biphenyl group to the flexible spacer is apparently less efficient at donating electrons than an amine group, and the fact that a considerable proportion of the complete polymer molecule will not be making any contribution to NLO activity will detract from $\chi^{(2)}$ if not β.

Series of side-chain molecules have been studied with different NLO active groups, side-chain lengths and backbones. The design requirements are a stable nematic phase for alignment and poling, the ability to quench to a glass avoiding strongly scattering smectic textures and long term stability of the glass. For the oxynitrobiphenyl material with a methacrylate backbone $\langle 78 \rangle$ (Fig. 8.21), the temperature range of liquid crystalline stability increases with increasing spacer length, in part as a consequence of a reduction in glass transition temperature from 84 °C for $-(\text{CH}_2)_3-$ to 10 °C for $-(\text{CH}_2)_{12}-$. However, longer spacers also increase the stability of smectic over nematic, and for $-(\text{CH}_2)_{12}-$ slow cooling can produce limited crystallinity. If the methacrylate backbone is replaced by a siloxane one $\langle 79 \rangle$ (Fig. 8.21), then for equivalent spacer lengths, the upper limit of mesophase stability is increased as the backbone interferes less with the side group order. On the other hand, the glass transition temperature of the siloxanes is much lower than the methacrylates. Commercial sources are understandably reticent regarding the values of $\chi^{(2)}$ which have been achieved with this type of material, although it is likely that they approach 10^{-8} esu. There is no reason why aligned side-chain polymers should not give respectable $\chi^{(3)}$ parameters, without of course the necessity for poling, although it seems likely that main-chain conjugated systems will give better performances in this respect.

(2) *Main chain LCPs.* The two most prominent types of linear molecule developed for their third order NLO properties are the polydiacetylenes ⟨68⟩ and the PBO and PBTZ family ⟨45, 46⟩. The polydiacetylenes are polymerized from single crystals of monomer and thus do not involve a liquid crystalline phase at any stage. The PBO/PBTZ types, however, can be oriented as lyotropic solutions in solvents such as methane sulphonic acid, techniques evolved originally for the production of ultra strong and stiff fibres. The highly oriented polymer, once the solvent is removed, gives values of $\chi^{(3)}$ of the order of 10^{-10} esu. It appears that even larger susceptibilities in excess of 10^{-9} esu with laser damage thresholds above 1 GW cm^{-2}, are achievable in heavily conjugated ladder polymers typified by that shown in Fig. 8.2(*a*) which, again, can best be aligned in a lyotropic mesophase. Ultimately however, other factors such as transparency and surface smoothness are likely to prove just as important in determining the practical efficiency of any NLO device.

At first sight a main-chain approach to second order NLO materials would not seem particularly practicable as long linear molecules need both a substantial dipole and sufficient kinetic freedom to be poleable. However, Levine and Bethea [8.32] demonstrated that molecules of poly(γ-benzyl-L-glutamate) (PBLG) ⟨22⟩, in solvents which stabilized the α helix form, could be effectively poled in comparatively modest electric fields. The polar nature of the molecule means that not only does it have a large dipole to facilitate orientation, but also an especially large value of β. Their value of 500×10^{-30} esu for high molecular weight polymer (500 000), is substantially larger than the coefficients

Fig. 8.21 Two side-chain liquid crystalline polymers developed for NLO activity.

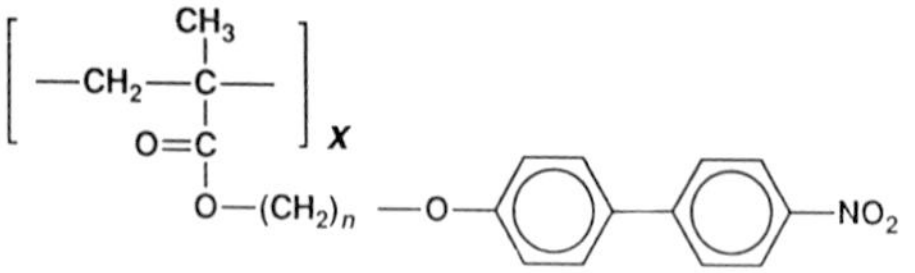

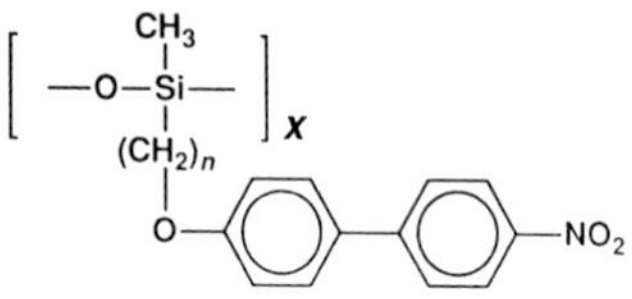

listed for the smaller side-chain groups of Table 8.3, and may be expected to correspond to $\chi^{(2)}$s above 10^{-7} esu.

(3) *Guest–host materials.* Another route by which comparatively small, NLO active, molecules can be aligned, poled and fabricated as a transparent solid is by incorporating them in a polymeric matrix. The host phase can be either liquid crystalline or isotropic, and it has been shown theoretically that the advantages of a mesophase matrix are equivalent to reducing the randomizing influence of temperature by a factor of three (Williams [8.33]). The quadrupolar orientation of the nematic matrix within the poling field assists the dipolar orientation of the NLO active guest molecules. One example of this approach has been described by Li *et al.* [8.34]. They dissolved molecules of an oligomeric fragment of **PBTZ**, which was terminated by opposing nitro and dimethyl amino groups, in the polymer itself. The resultant material had a commendable $\chi^{(2)}$ value of 6×10^{-9} esu. In order to maximize the efficiency of the guest–host approach, the concentration of the active guest molecules should be as large as solubility permits. However, once one is in the realm of concentrated solutions, it is possible to envisage a liquid crystalline phase stabilized by interactions between the guest molecules. Effectively, the phase is now lyotropic, and the advantages of making the active elements polymeric and removing the solvent, are clearly apparent.

8.4.4 *Types of NLO device*

Liquid crystalline polymers seem likely to play a key role in NLO device manufacture, and it is appropriate to review briefly the types of device which are now possible. In general, NLO behaviour enables the amplitude of a transmitted beam to be modified in some predetermined way as a function of its own input amplitude, or by an electric field applied to the medium. Such functionality opens up a wide range of possible device applications ranging from a straightforward electro-optic transducer, to a complex optical computer. The conception of new types of device is limited by the size of the NLO coefficients currently available, and in the area of optical–optical responses, sensitivities are yet to be achieved which permit the use of low powered solid state lasers as sources.

When light is shone through a second order non-linear material, some of the light emerges with twice the frequency, i.e. there is a colour change. The effect is particularly striking if the incident beam is in the near infra red, where the polymer appears to fluoresce in the invisible beam. The frequency doubling or second harmonic generation (SHG) properties not only provide a means of measuring $\chi^{(2)}$, but also offer other potential advantages. For example, the doubling of the frequency of semiconductor lasers can enhance the performance of photosensitive materials, and can also increase the capacity of optical data storage by a factor of perhaps four.

It is possible to modulate a beam of light passing through an NLO medium with an electric field at right angles to the optic axis. The device is known as a *Pockel cell,* and the field effectively changes the birefringence which controls the transmitted intensity. One great merit of organic NLO materials is that because their polarizability arises from electronic displacements rather than the ionic displacements which occur in inorganic materials, they are able to operate at much higher modulation frequencies for a given voltage. In the design of such devices it is important to remember that the optical path must ideally be at right angles to the poling axis so that the electric field is in the plane of the molecular director. Several working geometries are possible, one of which, shown in Fig. 8.22, has been described by Thackara *et al.* [8.35]. A side-chain polymer, with a flexible spacer of six units, was poled normal to the plane of the cell (Fig. 8.22) and the wave guided beam was thus favourably oriented for a good NLO response. It is of course possible to modulate transmitted light by field induced molecular reorientation effects, however the key advantage is in the response time. NLO effects are electronic and very fast, while molecular reorientation effects are many orders of magnitude slower. Figure 8.23 shows a comparison of the optical response of the NLO cell of Fig. 8.22, with that of a similar cell in which the active component was replaced by an unpoled small-molecule guest (methyl nitro-aniline)–polymeric host (PMMA) material, which depended on molecular reorientation for its activity. The differences in response at higher frequencies are most marked.

Fig. 8.22 Waveguide modulator using poled side-chain material. (After Thackara *et al.* [8.35].)

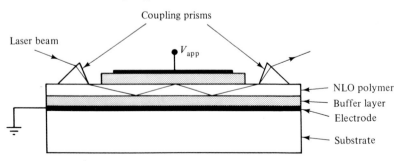

Whereas the difficulty of making film devices for transmission normal to the plane with second order materials is associated with achieving in-plane poling, the problem with third order materials is the size of the effect itself. With current $\chi^{(3)}$'s being of the order of 10^{-9} esu, which is indeed good for a non-resonant material, it is estimated (Lytel *et al.* [8.36]) that optical intensities of the order of several MW cm^{-2} would be required over path lengths of the order of 1 cm. So, the very considerable potential for optical–optical amplification and logic operations need to be considered within the context of long-path waveguide type geometries with the consequent requirement of good optical clarity. Currently, the power requirement per bit for $\chi^{(3)}$ polymers is about an order of magnitude greater than that for standard Ti–LiNbO$_3$ technology; on the other hand, the frequency response is also higher by at least the same margin.

8.5   Field-orientation devices in polymers

One objective in making electro-active polymeric liquid crystals is to achieve much of the functionality of small-molecule materials without the inherent drawback of having to contain a liquid phase within a cell. Side-chain materials provide the merits of a solid (or at least of a very high viscosity fluid) based on the properties of the

Fig. 8.23 Comparison of frequency response of a modulator depending on molecular reorientation (□) with one based on second order NLO activity (◆). (The loss at high frequency for the NLO material is associated with the limited response of the detection system.) (Replotted from Lytel *et al.* [8.36].)

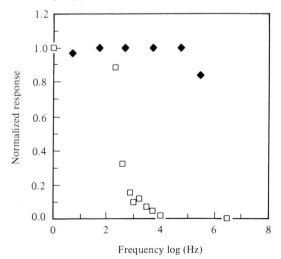

polymer backbone, while preserving some of the liquid phase mobility in the mesogenic groups despite their attachment to the backbone as side groups. It should be emphasized that most liquid crystal devices which depend on field induced molecular orientation for their activity can, in principle, be made to work with polymeric analogues. However, the move to the polymeric phase will involve higher elastic and viscous constants. In general, response times will be inconveniently long, although they can be controlled to some extent by adjusting the length of the flexible spacer connecting the side chain to the backbone. It would appear that the degree of molecular motion intrinsic to the operation of what is known as a chiral smectic C* device, may be compatible with a polymeric side-chain architecture, and such systems show promise in the quest for very large area (A4 page and above) displays without the need for active matrixing which involves the incorporation of thin film transistors within the display itself.

8.5.1   *Chiral smectic* (S_C^*) *devices*

Chiral smectic liquid crystals (see Chapter 2) provide the means of obtaining ferroelectric behaviour in a mesophase, in which there is long range orientation order with respect to both molecular alignment and polar sense. A useful property of this structure is that a bistable condition can be achieved with effective switching from one state to the other. The operation of a smectic S_C^* device is shown in Fig. 8.24. The incremental twist between successive tilted layers is undone or suppressed through the influence of surface fields. Both the normal to the smectic layers and the director lie in the plane of the surface, and the tilt of the director with respect to the layer normal can adopt orientations of $+\theta$ or $-\theta$. If the molecule has an effective electric dipole at a large angle to the axis of the rods, then this dipole will point away from the surface for one setting of θ and towards it for the opposite setting ($-\theta$). (The direction of these dipoles is shown by short arrows on the figure.) A change in the polarity of an electric field across the cell will change the direction of the electric dipole and effectively switch the molecular orientation between $\pm\theta$, the molecules precessing about the layer normal. Switching times are fast for a molecular reorientation process, and have been recorded at the 10 μs level. Such a phenomenon lends itself to polymeric liquid crystals in which the backbone is sufficiently flexible to ensure that the glass transition of the material is below ambient temperature. The mesogenic side chain must be designed to be chiral and will organise into tilted layers as in small-molecule materials. The backbone will, in general, remain between the layers, although there may be a degree of incorporation.

Fig. 8.24 The principle of operation of a S_C^* device. (*a*) A chiral smectic C with the helical twist quantized in the smectic layers; the director makes a constant angle θ with the helix axis. (*b*) Some of the director orientations on the surface of a cone (broken arrows). Assigned to each director is an electric dipole vector (solid arrows) which is here assumed to be normal to both the director and the helix axis. (*c*) The alignment of the helix axes parallel to an interface with a substrate causes the helix to lie flat so that there are only two possible director axes, both in the plane of the substrate at 2θ to each other. In one case the dipole vector will point away from the interface, in the other, towards it. (*d*) An external DC electric field normal to the plane of the film can be used to switch the dipole orientation and thus the director orientation, between the two states, each of which will remain stable when the field is removed. Note that C* phases can be made both from small-molecule and polymeric materials.

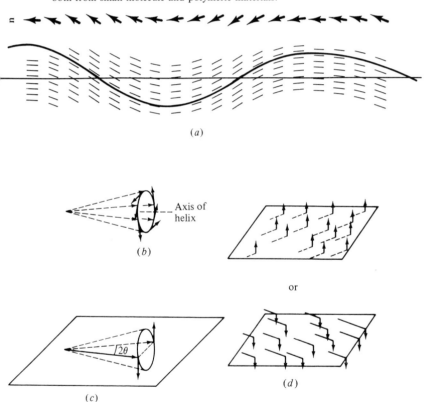

In conventional S_C^* liquid crystals the contrast between the two switchable states is obtained by viewing between crossed polars. In the case of S_C^* polymers, it is possible to graft on, as side groups, a proportion of dichroic dye or fluorescent molecules which will be incorporated within the smectic layers of the mesophase. In this way optical contrast can be obtained without the need for polars. While bistability is excellent for character writing and shutters, it does seem to rule out the possibility of grey levels, although intermediate contrast may be achieved through the use of time modulated bursts of pulses. An example of a side-chain liquid crystalline polymer showing a C* phase between 50 °C and 79 °C is shown as ⟨80⟩ in Fig. 8.25.

8.5.2   *Liquid crystalline elastomers*

The value of a polymeric backbone in greatly increasing the viscosity of a mesophase without preventing usefully rapid field orientation of the side groups has already been discussed. However, if a modest number of chemical cross links is introduced between the backbones, then the viscosity of the polymer will effectively become infinite to give rubbery or elastomeric properties. The result is that the material will sustain a stress field at a given level of mechanical deformation without viscous flow leading to complete relaxation. The exciting prospect from the device point of view is that a liquid crystalline elastomer enables stress field to be coupled into other fields through molecular orientation effects, and in particular to produce a very large stress–optical effect. A deviatoric stress field will induce quadrupolar alignment into any molecular entity, such as a side group, which is anisotropic in shape; while the fact that the groups are mutually aligned within the mesophase will greatly enhance the phenomenon by reducing the randomizing influence of thermal vibrations.

Molecules which give elastomeric mesophases at room temperature

Fig. 8.25 A liquid crystalline side-chain polymer which form a switchable chiral smectic, C*, phase. The chiral centre on the molecule is marked by *.

⟨80⟩

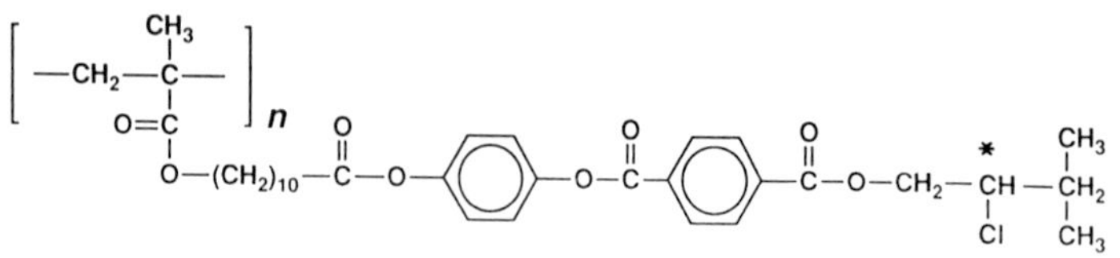

are normally based on siloxane backbones (Finkelmann and Rehage [8.37]), although some detailed studies have been made with acrylate backbones above their glass transition temperature (about 80 °C) by Mitchell *et al.* [8.38]. These latter polymers were made by copolymerizing the monomers shown in Fig. 8.26. Measurements of the elastic modulus of the rubber in the isotropic phase indicated that 2.8 % of the cross-linking monomer component gave an effective cross link density of about 1 %. Figure 8.27 is a plot of the orientation of the side groups as a function of the extensional strain in the elastomer, measured by X-ray diffraction after equilibrium had been reached at a

Fig. 8.26 Units which are copolymerized to form liquid crystalline elastomers. [8.38]

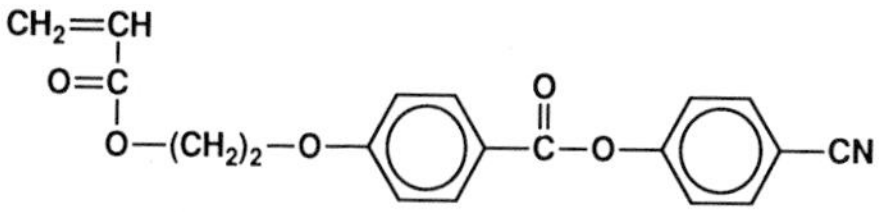

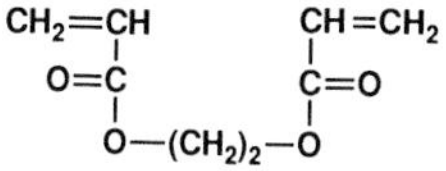

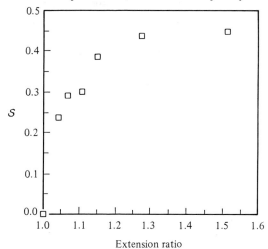

Fig. 8.27 Development of orientation of the side chains, expressed as the order parameter S, with strain in a liquid crystalline elastomer.

given applied strain. It is noteworthy that a large percentage of the total orientation is achieved at low strains.

The sensitivity of orientation to strain means that stress fields can be coupled to optical effects in much the same way as electric or magnetic fields. Furthermore, the addition of different active components as side chains on the same backbone opens up further possibilities of stress driven effects. For example dichroic dyes added as side chains to the same backbone, could enable direct reading (colour indicating) strain gauges to be made.

8.6   Laser-writable devices

This application depends on the sensitivity of microstructure to the local thermal history of the material. The additional levels of molecular organization associated with mesophases, and polymeric mesophases in particular, commend this type of material for laser-writable devices, although it should be emphasized that neither liquid crystallinity nor polymerization are always necessary to achieve such effects.

8.6.1   *Strongly scattering smectic textures*

Smectic mesophases are characterized by a texture known as 'focal conic' which has been described in Chapter 6. It is often formed on a fine scale, barely resolvable in an optical microscope, and in the case of polymeric materials the rate of coarsening is very slow indeed. The long term stability of such a microstructure and its high efficiency in scattering light commend it as the basis of a range of writable devices. The key fact is that the light scattering results from the variation, on a micron scale, in the orientation of elements of the material which are optically anisotropic. Hence, if the elements are aligned by the application of some external field, the strong scattering is lost and the material appears transparent. It is possible to write clear laser tracks on a scattering background, and scattering tracks on a clear background. It is also useful to be able to return the whole structure to its background state as this provides a means of erasing the written information.

The incorporation of dye molecules into smectic, laser-writable devices can of course be used to provide general colour, however they are probably more useful in increasing the efficiency of energy absorption from the laser beam, and can be matched to the laser frequency for maximum effect.

(1) *Scattering backgrounds.* A strongly scattering smectic polymer will provide a good background, as long as there is no significant

homeotropic alignment introduced by surface interactions. Non-scattering structures can be induced by the local heating of the laser in a variety of ways:

(a) the locally heated region can cool too rapidly for the strongly scattering smectic structure to reform. The track may be either isotropic, in which case the liquid crystallinity has been locally destroyed, or nematic.

(b) the smectic phase which reforms in the track is oriented, and thus transparent, either as a result of the electric field component of the beam or because of stresses associated with the contraction on cooling.

(c) the smectic phase in the track is oriented by the application of a general electric field to the sample during writing.

Erasure can generally be achieved by reheating the whole device. However, this is not normally convenient, and in some circumstances it is possible to recreate the scattering texture, by applying a high electric field at low frequency which effectively 'stirs' the structure. For polymeric smectics, where the glass transition temperature is above ambient, a component of general heating would still be necessary to take the material into the mesophase.

An example of a laser written track in a smectic liquid crystalline polymer is shown in Fig. 8.28. In this case the material in the track is aligned through the action of contraction stresses on cooling.

Fig. 8.28 Two clear laser tracks 'drawn' onto a strongly scattering smectic polymer. The larger track is 300 μm wide, the smaller, 3 μm. (Courtesy Dr H. Coles.)

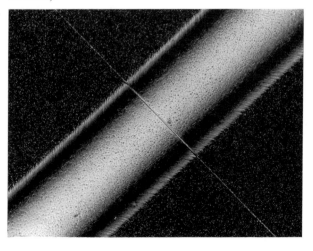

(2) *Clear backgrounds.* The complete area of a device can be rendered clear through either field or surface induced homeotropic alignment. Again, a modest field will usually suffice if it is applied during cooling, while a strong high frequency one (> 100 kHz) is necessary to produce the general alignment directly from the scattering smectic phase. The local heating associated with the laser beam without any other applied field, can produce a scattering smectic texture if the cooling is not too rapid to prevent the phase transformation. However, the other alignment factors in the track which were exploited for writing on a scattering background, must not be significant in this case. It is probably more reasonable to consider any scattering-on-clear device as still within the province of low molar mass smectic liquid crystals.

8.6.2   *Optically induced conformation transitions*

Some chromophore molecules related to dyes change their conformation in response to light. An example is the azo linkage where the absorption of energy in the visible range transforms the more stable *trans* state into the *cis* state (Fig. 8.29). This photo-induced isomerism is especially interesting where the chromophore is incorporated within a mesogenic group, for the application of light can have a direct influence on the stability of the liquid crystalline phase. Side-chain polymers with the mesogenic unit *trans* will obviously be more mesogenic than with the light induced *cis* kink. There is also evidence that changes can be induced both above and below the glass transition temperature [8.39, 8.40].

Fig. 8.29 Light activated conformation change.

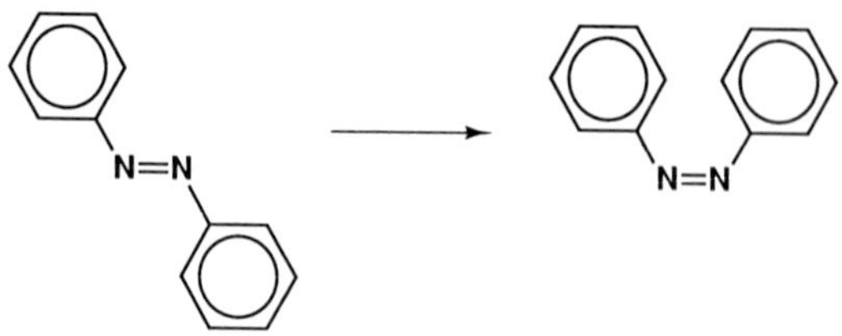

References

Chapter 1

1.1 Reinitzer, F. (1888) *Monatsch*, **9**, 421.

1.2 Lehmann, O. (1890) *Z. Phys. Chem.*, **5**, 427.

1.3 Friedel, G. (1922) *Ann. Physique*, **18**, 273.

1.4 Vorlander, D. (1923) *Z. Phys. Chem.*, **105**, 211.

1.5 Bawden, F. C. and Pirie, N. W. (1937) *Proc. Roy. Soc. London* Ser. B, **123**, 274.

1.6 Bernal, J. D. and Fanchuchen, I. (1941) *J. Gen. Physiol.*, **25**, 111.

1.7 Oster, G. (1950) *J. Gen. Physiol.*, **34**, 415.

1.8 Robinson, C. (1956) *Trans. Faraday. Soc.*, **52**, 571.

1.9 Ballard, D. G. H. (1958) Courtaulds Ltd, UK Patent BP 864, 962.

1.10 Kwolek, S. L. (1971) Du Pont, US Patent 3 600 350.

1.11 Jackson, W. J. and Kuhfuss, H. F. (1976) *J. Poly. Sci. Poly. Chem. Ed.*, **14**, 2043.

1.12 Onsager, L. (1949) *Ann. NY Acad. Sci.*, **51**, 627.

1.13 Flory, P. J. (1956) *Proc. Roy. Soc. London* Ser. A, **234**, 66.

1.14 Finkelmann, H., Happ, M., Portugall, M. and Ringsdorf, H. (1978) *Makromol. Chem.*, **179**, 2541.

1.15 McArdle, C. B. (ed.) (1989) *Side Chain Liquid Crystal Polymers*, Blackie, Glasgow.

Other historical reviews of the field:

Kelker, H. (1973) History of liquid crystals. *Mol. Cryst. Liq. Cryst.*, **21**, 1.

White, J. L. (1985) Historical survey of polymer liquid crystals. *J. Appl. Poly. Sci. Appl. Poly. Symp.*, **41**, 3.

Kelker, H. (1988) Survey of the early history of liquid crystals. *Mol. Cryst. Liq. Cryst.*, **165**, 1.

Jackson, W. J. (Jr.) (1989) Liquid crystal polymers XI. Liquid crystal aromatic polyesters: early history and future trends. *Mol. Cryst. Liq. Cryst.*, **169**, 23.

Percec, V. and Pugh, C. (1989) Molecular engineering of predominantly hydrocarbon-based LCPs, in *Side Chain Liquid Crystal Polymers*, ed. C. B. McArdle, Blackie, Glasgow, introduction.

Chapter 2

2.1 Elias, H-G. (1977) *Macromolecules*, Book 1, Wiley, London, Ch. 4.

2.2 Friedel, G. (1922) *Ann. Physique*, **18**, 273.

2.3 Gray, G. W. and Goodby, J. W. G. (1984) *Smectic Liquid Crystals*, Leonard Hill, Glasgow.

2.4 Goodby, J. W. G., Gray, G. W., Leadbetter, A. J. and Mazid, M. A. (1980) in *Liquid Crystals of One- and Two-Dimensional Order*, ed. W. Helfrich and G. Heppke, Springer series in Chemical Physics, Springer-Verlag, Berlin, Vol. 11, p. 3.

2.5 Sperling, L. H. (1986) *Introduction to Physical Polymer Science*, Wiley, New York.

2.6 Ward, I. M. (1983) *Mechanical Properties of Solid Polymers* (2nd edn), Wiley, Chichester, Ch. 2.

2.7 Boeffel, C. and Spiess, H-W. (1989) in *Side Chain Liquid Crystal Polymers*, ed. C. B. McArdle, Blackie, Glasgow, Ch. 8.

2.8 Lipson, S. G. and Lipson, H. (1981) *Optical Physics* (2nd edn), Cambridge University Press, Ch. 5.
2.9 Flory, P. (1953) *Principles of Polymer Chemistry*, Cornell University Press, Ithaca, Ch. 12.

Chapter 3
3.1 Chan, W-C., Mooney, J. A. and Windle, A. H. (1989) *Liquid Crystals* (in press).
3.2 Mooney, J. A. Unpublished data.
3.3 Hanna, S. and Windle, A. H. (1988) *Polym. Commun.*, **29**, 236.
3.4 Economy, J., Volken, W., Viney, C., Geiss, R., Siemens, R. and Kanig, T. (1988) *Macromols.*, **21**, 2777.
3.5 Hanna, S. and Windle, A. H. (1991) *Polymer* (in press).
3.6 Flory, P. J. and Ronca, G. (1979) *Mol. Cryst. Liq. Cryst.*, **54**, 311.
3.7 Ciferri, A. (1982) in *Polymer Liquid Crystals*, ed. A. Ciferri, W. R. Krigbaum and R. B. Meyer, Academic Press, New York.
3.8 Erdemir, A. B. (1982) *Fibres from Aromatic Copolyesters*, Ph.D. Thesis, Leeds University.
3.9 Gray, G. W. (1982) in *Polymer Liquid Crystals*, ed. A. Ciferri, W. R. Krigbaum and R. B. Meyer, Academic Press, New York.
3.10 Hummell, J. P. and Flory, P. J. (1980) *Macromols.*, **13**, 479.
3.11 Berger, K. and Ballauff, M. (1988) *Mol. Cryst. Liq. Cryst.*, **157**, 109.
3.12 Jackson, W. J. (1980) *Brit. Polym. J.*, **12**, 152.
3.13 US Patent belonging to Dartco Manufacturing Company (Dart-Kraft) 1984.
3.14 US Patent belonging to Hoechst Celanese Ltd. 1985.
3.15 Calundann, G. W. and Jaffe, M. (1982) *Proceedings of the Robert A. Welch Conference on Chemical Research. XXVI, Synthetic Polymers*, Houston, Texas, p. 247.
3.16 Jackson, W. J. and Kufuss, H. F. (1976) *J. Poly. Sci., Poly. Chem. Ed.*, **14**, 2043.
3.17 Flory, P. J. (1947) *J. Chem. Phys.*, **15**, 684.
3.18 Hanna, S. and Windle, A. H. (1988) *Polymer*, **29**, 207.
3.19 MacDonald, W. A. (1987) *Mol. Cryst. Liq. Cryst.*, **153**, 311.
3.20 Roviello, A. and Sirigu, A. (1982) *Makromol. Chem.*, **183**, 895.
3.21 Flory P. J. (1984) in *Advances in Polymer Science 59*, ed. M. Gordon, Springer-Verlag, Berlin, p. 1.
3.22 Miller, W. G., Wu, C. C., Wee, E. L., Santee, G. L., Ray, T. H. and Goebel, K. (1974) *Pure and Appl. Chem.*, **38**, 37.
3.23 Flory, P. J. (1956) *Proc. Roy. Soc.*, **234**A, 60, 73.
3.24 Nakajima, A., Hayashi, T. and Ohmori, M. (1968) *Biopolymers*, **6**, 973.
3.25 Hill, A. and Donald A. M. (1989) *Liq. Cryst.*, **6**, 93; Horton, J. and Donald, A. M. (1991) *Polymer* **32**, 2418.
3.26 Horton, J., Donald A. M. and Hill, A. (1990) *Nature*, **346**, 44.
3.27 Papkov, S. P. (1984) in *Advances in Polymer Sciences 59*, ed. M. Gordon, Springer-Verlag, Berlin, p. 75.
3.28 Fukuzawa, T., Uematsu, I. and Uematsu, Y. (1974) *Polymer J.*, **6**, 431.
3.29 Sasaki, S., Tokama, K. and Uematsu, I. (1983) *Polym. Bull.*, **10**, 539.
3.30 Wolfe, J. F. (1988) *Encyclo. Polymer Sci. and Eng.*, **11**, 601.
3.31 Hwang, W-F., Wiff, D. R., Verschoore, C., Price, G. E., Helminiac, T. E. and Adams, W. W. (1983) *Poly. Eng. Sci.*, **23**, 784.
3.32 Wiff, D. R., Helminiac, T. E. and Hwang, W-F. (1988) in *High Modulus Polymers*, ed. A. E. Zachariades and R. S. Porter, Dekker, New York.
3.33 Shibaev, V. P. and Plate, N. A. (1984) in *Advances in Polymer Science 60/61*, ed. M. Gordon, Springer-Verlag, Berlin.

3.34 Finkelmann, H. and Rehage, G. (1984) in *Advances in Polymer Science 60/61*, ed. M. Gordon, Springer-Verlag, Berlin.

3.35 Wang, X-J and Warner M. (1986) *J. Phys.*, **A19**, 2215.

3.36 Warner M. (1988) *Mol. Cryst. Liq. Cryst.*, **155**, 433.

Chapter 4

4.1 Onsager, L. (1949) *Ann. N.Y. Acad. Sci.*, **51**, 627.

4.2 Flory, P. J. (1956) *Proc. Roy. Soc.*, **234A**, 73.

4.3 de Gennes, P. G. (1974) *The Physics of Liquid Crystals*, Clarendon Press, Oxford, pp. 34ff.

4.4 Flory, P. J. (1984) in *Advances in Polymer Science 59*, ed. M. Gordon, Springer-Verlag, Berlin, p. 1.

4.5 Maier, W. and Saupe, A. (1959) *Z. Naturf.*, **14a**, 882.

4.6 Maier, W. and Saupe, A. (1960) *Z. Naturf.*, **15a**, 287.

4.7 Ishihara, A. (1951) *J. Chem. Phys.*, **19**, 1142.

4.8 Flory, P. J. and Ronca, G. (1979) *Mol. Cryst. Liq. Cryst.*, **54**, 289.

4.9 Miller, W. G. (1978) *Ann. Rev. Phys. Chem.*, **29**, 519.

4.10 Abe, A. and Flory, P. J. (1978) *Macromols.*, **11**, 1122.

4.11 Flory, P. J. and Frost, R. S. (1978) *Macromols.*, **11**, 1126.

4.12 Frost, R. S. and Flory, P. J. (1978) *Macromols.*, **11**, 1134.

4.13 Flory, P. J. (1953) *Principles of Polymer Chemistry*, Cornell University Press, Ithaca, pp. 321ff.

4.14 Wee, E. L. and Miller, W. G. (1978) in *Liquid Crystals and Ordered Fluids*, vol 3, eds. J. F. Johnson and R. S. Porter, Plenum, New York, p. 371.

4.15 Balbi, C., Bianchi, E., Ciferri, A. and Tealdi, A. (1980) *J. Poly. Sci. Phys.*, **18**, 2037.

4.16 Bair, T. I., Morgan, P. W. and Kilian, F. L. (1977) *Macromols.*, **10**, 1396.

4.17 Conio, G., Bianchi, E., Ciferri, A. and Tealdi, A. (1981) *Macromols.*, **14**, 1084.

4.18 Aharoni, S. M. (1983) *Polym. Bull.*, **9**, 186.

4.19 Chandrasekhar, S. and Madhusudana, N. V. (1972) *Mol. Cryst. Liq. Cryst.*, **17**, 37.

4.20 Warner, M. and Flory, P. J. (1980) *J. Chem. Phys.*, **73**, 6327.

4.21 Kratky, O. and Porod, G. (1949) *Rec. Trav. Chim.*, **68**, 1106.

4.22 Papkov, S. P. (1977) *J. Poly. Sci. USSR*, **19**, 1.

4.23 Flory, P. J. (1978) *Macromols.*, **11**, 1138.

4.24 Ten Bosch, A., Maissa, P. and Sixou, P. (1983) *Phys. Lett.*, **94A**, 298.

4.25 Ten Bosch, A., Maissa, P. and Sixou, P. (1985) *Polymer Liquid Crystals*, ed. A. Blumstein, Plenum, New York, p. 109.

4.26 Ronca, G. and Yoon, D. Y. (1982) *J. Chem. Phys.*, **76**, 3295.

4.27 Pincus, P. and De Gennes, P. G. (1978) *J. Poly. Sci. Poly. Symp.*, **65**, 85.

4.28 Khoklov, A. R. and Semenov, A. N. (1981) *Physica*, **108A**, 546.

4.29 Corradini, P. and Vacatello, M. (1983) *Mol. Cryst. Liq. Cryst.*, **97**, 119.

4.30 Matheson, R. R. and Flory, P. J. (1981) *Macromols.*, **14**, 954.

4.31 Vasilenko, S. V., Khokhlov, A. R. and Shibaev, V. P. (1984) *Macromols.*, **17**, 2270.

4.32 Poland, D. and Scheraga, H. A. (1970) *Theory of the Helix–Coil Transition in Biopolymers*, Academic Press, New York.

4.33 Norisuye, T., Teramoto, A. and Fujita, H. (1973) *Polymer J.*, **4**, 323.

4.34 Kim, J. R. and Ree, T. (1984) *Polymer J.*, **16**, 669.

4.35 Matheson, R. R. (1983) *Biopolymers*, **22**, 43.

4.36 Robinson, C., Ward, J. C. and Beevers, R. B. (1958) *Faraday Discuss. Chem. Soc.*, **25**, 29.

4.37 Ackermann, T. and Ruterjans, H. (1964) *Z. Phys. Chem.*, **41**, 116.

4.38 Frenkel, S. Y., Shaltyka, L. G. and Elyashevich, G. K. (1970) *J. Poly. Sci.*, **C30**, 47.
4.39 Subramanian, R., Wittebolt, R. J. and Du Pré, D. P. (1982) *J. Chem. Phys.*, **77**, 4694.
4.40 Warner, M. (1989) in *Side Chain Liquid Crystal Polymers*, ed. C. B. McArdle, Blackie, Glasgow, p. 7.
4.41 Wang, X.-J. and Warner, M. (1987) *J. Phys.*, **A20**, 713.
4.42 Kirste, R. G. and Ohm, H. G. (1985) *Makromol. Chem., Rapid Comm.*, **6**, 179.
4.43 Keller, P., Carvalho, B., Cotton, J. P., Lambert, M., Moussa, F. and Pepy, G. (1985) *J. de. Phys. Letts.*, **46**, L1065.
4.44 Renz, W. and Warner, M. (1986) *Phys. Rev. Lett.*, **56**, 1268.

Chapter 5
5.1 Friedel, G. (1922) *Ann. Phys.*, **18**, 273.
5.2 Ronca, G. and Yoon, D. Y. (1982) *J. Chem. Phys.*, **76**, 3295.
5.3 Noel, C., Laupretre, F., Friedrich, C., Fayolle, B. and Bosio, L. (1984) *Polymer*, **25**, 263.
5.4 Millaud, B., Thierry, A. and Skoulios, A. (1978) *Mol. Cryst. Liq. Cryst.*, *(Lett.)*, **41**, 263.
5.5 Wahlstrom, E. E. (1960) *Optical Crystallography*, Wiley, New York.
5.6 Windle, A. H., Viney, C., Golombok, R., Donald, A. M. and Mitchell, G. R. (1985) *Faraday Discuss. Chem. Soc.*, **79**, 55ff and 102.
5.7 Viney, C., Mitchell, G. R. and Windle, A. H. (1985) *Mol. Cryst. Liq. Cryst.*, **129**, 75.
5.8 Sawyer, L. C. and Jaffe, M. (1986) *J. Materials Science*, **21**, 1897.
5.9 Finkelmann, H. (1982) in *Polymer Liquid Crystals*, ed. A. Ciferri, W. R. Krigbaum and R. B. Meyer, Academic Press, New York.
5.10 Kirste, R. G. and Ohm, H. G. (1985) *Makromol. Chem., Rapid Comm.*, **6**, 179.
5.11 Hessel, F. and Finkelmann, H. (1986) *Polym. Bull.*, **15**, 349.
5.12 Uematsu, I. and Uematsu, Y. (1984) in *Advances in Polymer Science 59*, ed. M. Gordon, Springer-Verlag, Berlin, p. 37.
5.13 Keating, P. N. (1969) *Mol. Cryst. Liq. Cryst.*, **8**, 315.
5.14 Finkelmann, H., Ringsdorf, H., Siol, W. and Wendorff, J. H. (1978) *Makromol. Chem.*, **179**, 829.
5.15 Finkelmann, H. and Rehage, G. (1980) *Makromol. Chem., Rapid Comm.*, **1**, 733.
5.16 Ritter, A. P. (1986) Unpublished work.
5.17 Grandjean, F. (1921). *C. R. Hebd. Séan Acad. Sci.*, **172**, 71.
5.18 de Vries, H. (1951) *Acta Crystall.*, **4**, 219.
5.19 Shibaev, V. P. and Platé, N. A. (1984) in *Advances in Polymer Science 60/61*, ed. M. Gordon, Springer-Verlag, Berlin, p. 173.
5.20 Zugenmaier, P. and Mugge, J. (1984) *Makromol. Chem., Rapid Comm.*, **5**, 11.
5.21 Berger, K. and Ballauff, M. (1988) *Mol. Cryst. Liq. Cryst.*, **157**, 109.
5.22 Duran, R., Guillon, D., Gramain, P. and Skoulios, A. (1987) *Makromol. Chem., Rapid Comm.*, **8**, 181, (and) 321.
5.23 Herrmann-Schonherr, O., Wendorff, J. H., Ringsdorf, H. and Tschirner, P. (1986) *Makromol. Chem., Rapid Comm.*, **7**, 791.
5.24 Ringsdorf, H., Tschirner, P., Herrmann-Schonherr, O. and Wendorff, J. H. (1987) *Makromol. Chem.*, **188**, 1431.
5.25 Ebert, M., Herrmann-Schonherr, O., Wendorff, J. H., Ringsdorf, H. and Tschirner, P. (1988) *Makromol. Chem., Rapid Comm.*, **9**, 445.
5.26 de Vries, H. (1969) *Acta Crystall.*, **A25**, S135.

5.27 Blumstein, A., Vilasagar, S., Ponrathnam, S., Clough, S. B., Blumstein, R. B. and Maret, G. (1982) *J. Poly. Sci, Poly. Phys.*, **20**, 877.
5.28 Stewart, G. W. and Morrow, R. M. (1927) *Phys. Rev.*, **30**, 232.
5.29 Atkins, E. D. T. and Thomas, E. L. Private communication (to be published).
5.30 Lenz, R. W. (1985) *Faraday Discuss. Chem. Soc.*, **79**, 21.
5.31 Blumstein, A., Gauthier, M. M., Thomas, O. and Blumstein, R. B. (1985) *Faraday Discuss. Chem. Soc.*, **79**, 33.
5.32 Suzuki, T., Okawa, T., Ohnuma, T. and Sakon, Y. (1988) *Makromol. Chem., Rapid. Comm.*, **9**, 755.
5.33 Dobb, M. G., Johnson, D. J. and Saville, B. P. (1977) *J. Poly. Sci., Poly. Phys.*, **15**, 2201.
5.34 Dobb, M. G., Johnson, D. J. and Saville, B. P. (1979) *Polymer*, **20**, 1284.
5.35 Hanna, S. (1989) Ph.D. Thesis, Cambridge University.
5.36 Spontak, R. J. and Windle, A. H. (1990) *J. Mat. Sci.*, **25**, 2727.
5.37 Gutierrez, G. A., Chivers, R. A., Blackwell, J., Stamatoff, J. B. and Yoon, H. (1983) *Polymer*, **24**, 937.
5.38 Schrödinger, E. (1944) *What is Life?*, Cambridge University Press.
5.39 Hanna, S. and Windle, A. H. (1988) *Polymer*, **29**, 207.

Chapter 6

6.1 Donald, A. M. and Windle, A. H. (1985) in *Recent Advances in Liquid Crystalline Polymers*, ed. L. L. Chapoy, Elsevier, London, Ch. 12.
6.2 De Gennes, P. G. (1977) *Mol. Cryst. Liq. Cryst. (Lett.)* **34**, 177.
6.3 Millaud, B., Thierry, A. and Skoulios, A. (1978) *J. de Phys.*, **39**, 1109.
6.4 Meyer, R. B. (1985) *Faraday Discuss. Chem. Soc.*, **79**, 125.
6.5 De Gennes, P. G. (1968) *Solid St. Comm.*, **6**, 168.
6.6 Meyer, R. B. (1968) *Appl. Phys. Lett.*, **14**, 208.
6.7 Du Pré, D. B. and Duke, R. W. (1975) *J. Chem. Phys.*, **63**, 143.
6.8 Freedericksz, V. (Frederiks) and Zolina, V. (1933) *Trans. Faraday Society*, **29**, 919.
6.9 Freedericksz, V. (Frederiks) and Zwetkoff, V. (1934) *Sov. Phys.*, **6**, 490.
6.10 Sun, Z-M. and Klèman, M. (1984) *Mol. Cryst. Liq. Cryst.*, **111**, 321.
6.11 Fernandes, J. R. and Du Pré, D. B. (1981) *Mol. Cryst. Liq. Cryst. (Lett.)*, **72**, 67.
6.12 Fernandes, J. R. and Du Pré, D. B. (1984) in *Liquid Crystals and Ordered Fluids vol. 4*, eds. A. C. Griffin and J. F. Johnson, Plenum, New York, Ch. 18.
6.13 Meyer, R. B. (1983) in *Polymer Liquid Crystals*, eds. A. Ciferri, W. R. Krigbaum and R. B. Meyer, Academic Press, New York. Ch. 6.
6.14 Chandrasekhar, S. (1977) *Liquid Crystals*, Cambridge University Press.
6.15 Viney, C. and Windle, A. H. (1982) *J. Mat. Sci.*, **17**, 2661.
6.16 Meyer, R. B. (1973) *Phil. Mag.*, **27**, 405.
6.17 Cladis, P. and Klèman, M. (1972) *J. de Phys.*, **33**, 591.
6.18 Klèman, M. (1983) *Points Lines and Walls*, Wiley, Chichester.
6.19 Klèman, M., Liebert, L. and Strezelecki, L. (1983) *Polymer*, **24**, 295.
6.20 Donald, A. M., Viney, C. and Windle, A. H. (1985) *Phil. Mag.*, **52B**, 925.
6.21 Donald, A. M. and Windle, A. H. (1984) *J. Mat. Sci.*, **19**, 2085; (1984) *Polymer*, **25**, 1235.
6.22 Millaud, B., Thierry, A. and Skoulios, A. (1979) *J. de Phys. Lett.*, **40**, L-607.
6.23 Mazelet, G. and Klèman, M. (1986) *Polymer*, **27**, 714.
6.24 Hudson, S. D. and Thomas, E. L. (1989) *Phys. Rev. Lett.*, **62**, 1993.
6.25 Bouligand, Y. (1972) *J. de Phys.*, **33**, 525, 715.

6.26 Chiellini, E. and Galli, G. (1985) in *Recent Advances in Liquid Crystalline Polymers*, ed. L. L. Chapoy, Elsevier, London, p. 15.
6.27 Thomas, E. L. and Wood, B. A. (1985) *Faraday Discuss. Chem. Soc.*, **79**, 229.
6.28 Voigt-Martin, I. G., Durst, H., Reck, B. and Ringsdorf, H. (1988) *Macromols.*, **21**, 1620.
6.29 Hartshorne, N. H. and Stuart, A. (1970) *Crystals and the Polarising Microscope* (2nd edn), Arnold, London.
6.30 Gray, W. W. and Goodby, J. W. (1983) *Smectic Liquid Crystals*, Leonard Hill, Glasgow, Ch. 1.
6.31 Friedel, G. (1922) *Ann. Physique*, **18**, 273.
6.32 Krigbaum, W. R. and Watanabe, J. (1983) *Polymer*, **24**, 1299.
6.33 Noel, C. (1985) in *Recent Advances in Liquid Crystalline Polymers*, ed. L. L. Chapoy, Elsevier, London, Ch. 1.
6.34 Finkelmann, H. and Rehage, G. (1984) in *Advances in Polymer Science 60/61*, ed. M. Gordon, Springer-Verlag, Berlin, p. 99.
6.35 Shibaev, V. P. and Platé, N. A. (1984) in *Advances in Polymer Science 60/61*, ed. M. Gordon, Springer-Verlag, Berlin, p. 173.
6.36 Martin, P. G. and Stupp, S. I. (1988) *Macromols.*, **21**, 1222.
6.37 Martin, P. G. and Stupp, S. I. (1988) *Macromols.*, **21**, 1228.
6.38 Meurisse, P., Noel, C., Monnerie, L. and Fayolle, B. (1981) *Brit. Poly. J.*, **13**, 55.
6.39 Marrucci, G. (1984) *Int. Congr. Rheology Acapulco*, 1984.
6.40 Wissbrun, K. F. (1985) *Faraday Discuss. Chem. Soc.*, **79**, 161.
6.41 Donald A. M. and Windle, A. H. (1985) *Polymer*, **25**, 1235.
6.42 General discussion in *Faraday Discuss. Chem. Soc.* (1985) **79**, 274.

Chapter 7

7.1 Meyer, R. B. (1982) in *Polymer Liquid Crystals*, ed. A. Ciferri, W. R. Krigbaum and R. B. Meyer, Academic Press, New York, p. 133.
7.2 Donald, A. M. and Windle, A. H. (1984) *J. Mat. Sci.*, **19**, 2085; Donald, A. M. (1986) *Polymer Comm.*, **27**, 18.
7.3 Hermans, J. (1962) *J. Coll. Sci.*, **17**, 638.
7.4 Baird, D. G. (1978) in *Liquid Crystalline Order in Polymers*, ed. A. Blumstein, Academic Press, New York, Ch. 7.
7.5 Wissbrun, K. F. (1980) *Brit. Poly. J.*, **12**, 163; Wissbrun, K. F. (1981) *J. Rheol.*, **25**, 619.
7.6 Doi, M. (1975) *J. de Phys.*, **36**, 607.
7.7 Asada, T., Muramatsu, H., Watanabe, R. and Onogi, S. (1980) *Macromols.*, **13**, 867; Asada, T. (1982) in *Polymer Liquid Crystals*, ed. A. Ciferri, W. R. Krigbaum and R. B. Meyer, Academic Press, New York.
7.8 Ericksen, J. L. (1960) *Arch. Ration. Mech. Anal.*, **4**, 231.
7.9 Leslie, F. M. (1966) *Quart. J. Mech. Appl. Math.*, **19**, 357.
7.10 De Gennes, P. G. (1974) *The Physics of Liquid Crystals*, Oxford University Press, p. 153.
7.11 Miesowicz, M. (1946) *Nature*, **158**, 27.
7.12 Berry, G. C. (1988) *Mol. Cryst. Liq. Cryst.*, **165**, 333.
7.13 Pieranski, P., Brochard, F. and Guyon, E. (1973) *J. de Phys.*, **34**, 35.
7.14 Martins, A. F., Esnault, P. and Volino, F. (1986) *Phys. Rev. Lett.*, **57**, 1745.
7.15 Taratuta, V. G., Hurd, A. J. and Meyer, R. B. (1985) *Phys. Rev. Lett.*, **55**, 246.
7.16 Se, K. and Berry, G. (1987) *Mol. Cryst. Liq. Cryst.*, **153**, 133.

7.17 Casagrande, C., Fabre, P., Veyssiè, M., Weill, C. and Finkelmann, H. (1984) *Mol. Cryst. Liq. Cryst.*, **113**, 193; Fabre, P., Casagrande, C., Veyssiè, M. and Finkelmann, H. (1984) *Phys. Rev. Lett.*, **53**, 993.

7.18 Platé, N. A. and Shibaev, V. P. (1987) in *Comb-Shaped Polymers and Liquid Crystals*, Plenum, New York, p. 324.

7.19 Doi, M. and Edwards, S. F. (1986) *The Theory of Polymer Dynamics*, Oxford University Press, Ch. 10.

7.20 Doi, M. (1980) *Ferroelectrics*, **30**, 247; Doi, M. (1982) *J. Poly. Sci. Phys.*, **20**, 1963.

7.21 Berry, G. C., Se, K. and Srinivasrao, M. (1988) in *High Modulus Polymers*, ed. A. E. Zachariades, and R. S. Porter, Dekker, New York, p. 195.

7.22 Odell, J. A., Keller, A. and Atkins, E. D. T. (1985) *Macromols.*, **18**, 1443.

7.23 Wunder, S. L., Ramachandran, S., Gochanour, C. R. and Weinberg, M. (1986) *Macromols.*, **19**, 1696.

7.24 Onogi, S. and Asada, T. (1980) in *Rheology*, vol. 1, ed. G. Astarita, G. Marrucci and L. Nicholais, Plenum, New York.

7.25 Alderman, N. J. and Mackley, M. R. (1985) *Faraday Discuss. Chem. Soc.*, **79**, 149.

7.26 Nicholson, T. M. (1988) Ph.D. Thesis, Cambridge University.

7.27 Berry, G. C., Se, K. and Srinisvasrao, M. (1988) in *High Modulus Polymers*, ed. A. E. Zachariades and R. S. Porter, Dekker, New York, p. 195.

7.28 Kulichikhin, V. G. (1989) *Mol. Cryst. Liq. Cryst.*, **169**, 51.

7.29 Calundann, G., Jaffe, M., Jones, R. S. and Yoon, H. (1988) in *Fibre Reinforcements for Composite Materials*, ed. A. R. Bunsell, Elsevier, Amsterdam, Ch. 5.

7.30 Lewis, D. N. and Fellers, J. F. (1988) in *High Modulus Polymers*, eds. A. E. Zachariades and R. S. Porter, Dekker, New York, p. 1.

7.31 Dobb, M. G., Johnson, D. J. and Saville, B. P. (1977) *J. Poly. Sci. Phys.*, **15**, 2201.

7.32 Viney, C., Donald, A. M. and Windle, A. H. (1983) *J. Mat. Sci.*, **18**, 1136.

7.33 Haase, W. (1989) in *Side Chain Liquid Crystal Polymers*, ed. C. B. McArdle, Blackie, Glasgow, p. 309.

7.34 Finkelmann, H. and Rehage, G. (1984) in *Advances in Polymer Science 60/61*, ed. M. Gordon, Springer-Verlag, Berlin, p. 99.

7.35 Attard, G. S., Araki, K. and Williams, G. (1987) *J. Mol. Electron.*, **3**, 1.

7.36 Iizuka, E (1976) in *Advances in Polymer Science 20*, ed. M. Gordon, Springer-Verlag, Berlin, p. 79.

7.37 Iizuka, E. (1985) *J. Appl. Poly. Sci., Appl. Poly. Symp.*, **41**, 131.

7.38 Martin, P. G. and Stupp, S. I. (1987) *Polymer*, **28**, 897.

7.39 Haws, C. M., Clark, M. G. and Attard, G. S. (1989) in *Side Chain Liquid Crystal Polymers*, ed. C. B. McArdle, Blackie, Glasgow, p. 196.

7.40 Kol'tsov, L. G., Bel'nikevich, N. G., Gribanov, A. V., Papkov, S. P. and Frenkel, S. Y. (1973) *Vysokomol Soyed Ser B*, **15**, 645.

7.41 Moore, R. C. and Denn, M. M. (1988) in *High Modulus Polymers*, ed. A. E. Zachariades and R. S. Porter, Dekker, New York, p. 169.

7.42 Hardouin, F., Archard, M. F., Gasparoux, H., Liebert, L. and Strzelecki, L. (1982) *J. Poly. Sci. Phys.*, **20**, 975.

7.43 Maret, G. and Blumstein, A. (1982) *Mol. Cryst. Liq. Cryst.*, **88**, 295.

7.44 Anwer, A. and Windle, A. H. (1991) *Polymer*, **32**, 103.

7.45 Dubois-Violette, E., Durand, G., Guyon, E., Manneville, P. and Pieranski, P. (1978) in *Liquid Crystals*, ed. L. Liebert, Academic Press, New York, p. 147.

7.46 Chandrasekhar, S. and Kini, U. D. (1982) in *Polymer Liquid Crystals*, ed. A. Ciferri, W. R. Krigbaum and R. B. Meyer, Academic Press, New York, p. 201.

7.47 Lonberg, F., Fraden, S., Hurd, A. J. and Meyer, R. B. (1984) *Phys. Rev. Lett.*, **52**, 1903; Lonberg, F. and Meyer, R. B. (1985) *Phys. Rev. Lett.*, **55**, 718; Hurd, A. J., Fraden, S., Lonberg, F. and Meyer, R. B. (1985) *J. de Phys.*, **46**, 905.
7.48 Krigbaum, W. R. (1982) in *Polymer Liquid Crystals*, ed. A. Ciferri, W. R. Krigbaum and R. B. Meyer, Academic Press, New York, p. 275.

Chapter 8
8.1 Klei, H. E. and Stewart, J. J. P. (1986) *Int. J. Quantum. Chem.*, **20**, 529.
8.2 Sakurada, I. and Kaji, K. (1970) *J. Pol. Sci.*, **31**, 57.
8.3 Capaccio, G., Gibson, A. G. and Ward, I. M. (1979) in *Ultra High Modulus Polymers*, ed. A. Ciferri and I. M. Ward, Applied Sci., London, Ch. 1.
8.4 Calundann, G., Jaffee, M., Jones, R. S. and Yoon, H. (1988) in *Fibre Reinforcements for Composite Materials*, ed. A. R. Bunsell, Elsevier, Amsterdam, Ch. 5.
8.5 Lemstra, P. J., Kirschbaum, R., Ohta, T. and Yasuda, A. (1987) in *Developments in Oriented Polymers 2*, ed. I. M. Ward, Elsevier Appl. Sci., London and New York, Ch. 2.
8.6 Burkert, U. and Allinger, N. L. (1982) *American Chemical Society Monograph, 177*, Washington, DC.
8.7 Sorensen, R. A., Liau, W. B., Kesner, L. and Boyd, R. H. (1988) *Macromolecules*, **21**, 200.
8.8 Weirschke, S. G. (1988) M.Sc. Thesis, University of Dayton, Dayton, Ohio, USA.
8.9 Baughman, R. H., Gleiter, H. and Sendfeld, N. (1975) *J. Poly. Sci. Poly. Phys. Ed.*, **13**, 1871.
8.10 Galiotis, C. and Young, R. J. (1983) *Polymer*, **24**, 1023.
8.11 Davies, G. R. and Ward, I. M. (1988) in *High Modulus Polymers*, ed. A. E. Zachariades and R. S. Porter, Dekker, New York, Ch. 2.
8.12 Blundell, D. J. and Buckingham, K. A. (1985) *Polymer*, **26**, 1623.
8.13 Clements, J., Humphreys, J. and Ward, I. M. (1986) *J. Poly. Sci., Poly. Phys. Ed.*, **24**, 2293.
8.14 Kausch, H. H. (1978) *Polymer Fracture*, Springer-Verlag, Berlin.
8.15 Schaefgen, J. R., Bair, T. I., Ballou, J. W., Kwolek, S. L., Morgan, P. W., Panar, M. and Zimmerman, J. (1979) in *Ultra High Modulus Fibres*, ed. A. Ciferri and I. M. Ward, Applied Sci., London.
8.16 Jackson, Jr., W. J. and Kuhfuss, H. F. (1976) *J. Poly. Sci., Poly. Chem.*, **14**, 2043.
8.17 Termonia, Y., Meakin, P. and Smith, P. (1985) *Macromols.*, **18**, 2246; (1986) *Macromols.*, **19**, 154.
8.18 Termonia, Y., Meakin, P. and Smith, P. (1986) *Macromols.*, **19**, 154.
8.19 Termonia, Y. and Smith, P. (1988) in *High Modulus Polymers*, ed. A. E. Zachariades and R. S. Porter, Dekker, New York, Ch. 11.
8.20 Adams, W. W. and Eby, R. K. (1987) *MRS Bulletin*, (November), 22.
8.21 Troughton, M. J., Davies, G. R. and Ward, I. M. (1989) *Polymer*, **30**, 58.
8.22 Garg, S. K. and Kenig, S. (1988) in *High Modulus Polymers*, ed. A. E. Zachariades and R. S. Porter, Dekker, New York, Ch. 3.
8.23 Yamamoto, F. (1987) *Mol. Cryst. Liq. Cryst.*, **152**, 423.
8.24 Cox, M. K. (1987) *Mol. Cryst. Liq. Cryst.*, **152**, 415.
8.25 Burroughes, J. and Friend, R. (1988) *Physics World*, **1**, 24.
8.26 Mohlmann, G. R. and van der Vorst, C. P. J. M. (1989) in *Side Chain Liquid Crystal Polymers*, ed. C. B. McArdle, Blackie, Glasgow, Ch. 12.
8.27 Stamatoff, J. B., Buckley, A., Calundann, G., Choe, W. W., DeMartino, R.,

Khanarian, G., Leslie, T., Nelson, G., Stuetz, D., Teng, C. C. and Yoon, H. N. (1987) *Proceedings of SPIE Int. Soc. Opt. Eng.*, **682**, 85.

8.28 DeMartino, R., Choe, E. W., Khanarian, G., Hass, D., Leslie, T., Nelson, G., Stamatoff, J. B., Stuetz, D., Teng, C. C. and Yoon, H. N. (1988) in *Nonlinear Optical and Electroactive Polymers*, ed. P. N. Prasad and D. R. Ulrich, Plenum, New York.

8.29 Medrano, J. M. (1989) Symposium *Materials Science and Engineering of Rigid Rod Polymers*, Materials Research Society, Proceedings, volume 134.

8.30 Prasad, P. N. (1988) in *Nonlinear Optical and Electroactive Polymers*, ed. P. N. Prasad and D. R. Ulrich, Plenum, New York.

8.31 Ulrich, D. R. (1988) in *Nonlinear Optical and Electroactive Polymers*, ed. P. N. Prasad and D. R. Ulrich, Plenum, New York.

8.32 Levine, B. F. and Bethea, C. G. (1976) *J. Chem. Phys.*, **65**, 1989.

8.33 Williams, D. J. (1987) in *Nonlinear Optical Properties of Organic Molecules*, vol. 1, ed. D. S. Chemla and J. Zyss, Academic Press, Orlando.

8.34 Li, D., Marks, T. J. and Ratner, M. A. (1986) *Chem. Phys. Lett.*, **131**, 370.

8.35 Thackara, J. I., Lipscomb, G. F., Lytel, R., Stiller, M., Okasaki, E., DeMartino, R. and Yoon, H. (1987) *Proc. Conference on Lasers and Electrooptics, OSA*, Washington, DC.

8.36 Lytel, R., Lipscomb, G. F., Thackara, J., Altman, J., Elizondo, P., Stiller, M. and Sullivan, B. (1988) in *Nonlinear Optical and Electroactive Polymers*, ed. P. N. Prasad and D. R. Ulrich, Plenum, New York.

8.37 Finkelmann, H. and Rehage, G. (1984) in *Advances in Polymer Science 60/61*, ed. M. Gordon, Springer-Verlag, Berlin.

8.38 Mitchell, G. R., Davis, F. and Ashman, A. (1987) *Polymer*, **28**, 639.

8.39 Eich, M. and Wendorff, J. H. (1987) *Makromol. Chem., Rapid Comm.*, **8**, 467.

8.40 Reck, B. and Ringsdorf, H. (1987) *Makromol. Chem., Rapid Comm.*, **8**, 59.

Symbol index

A cross sectional area

$\mathscr{A}$ free energy

$\mathbf{A}$ shear strain rate tensor

A_j number of lattice sites available for first unit of the first sequence of the rod in the Flory model

$\mathscr{B}$ bending constant

$\mathbf{B}$ magnetic field

b extrapolation length

$\mathscr{C}$ constant of interaction energy in the Maier–Saupe theory

c concentration

$\mathbf{D}$ electric field displacement vector

D_r rotational diffusion coefficient

D_{ro} free rotational diffusion coefficient in dilute solution

D_{to} free translational diffusion coefficient

d molecular diameter

E applied field

E axial (Young's) modulus of a fibre

$\mathbf{E}$ electric field vector

F force

G shear modulus

G' storage modulus

G'' loss modulus

$G(\mathbf{R}, \mathbf{R}', N)$ probability of chain conformation starting at $\mathbf{R}$ and ending at $\mathbf{R}''$ with N elements

H enthalpy

$\mathbf{H}$ magnetic field

H_c critical magnetic field strength

j index for a molecule in the Flory model

K elastic (Frank) constant

k Boltzmann's constant

$\mathbf{k}$ wave vector

L contour length, rod length

L_c cut-off length

l_k Kuhn step length

l_0 link length

M molecular weight

M_n number average molecular weight

M_w	weight average molecular weight
m	number of decoupling units between backbone and mesogenic side-chain unit (length of flexible spacer)
N	total number of lattice sites in the Flory model
N_A	Avogadro's number
N_j	probability that second (and subsequent) sites of each sequence are also vacant in the Flory model
n	director
n	number of links/moles/degree of polymerization
n	refractive index
n_1, n_2, n_3	principal refractive indices
n_k	number of steps in worm-like chain
n_p	number of identical rods on lattice in the Flory model
n_{py}	total number of rods with misorientation y in the Flory model
n_r	radial refractive index
n_s	number of lattice sites occupied by solvent molecules in the Flory model
n_θ	azimuthal refractive index
P	polarizability
$\mathscr{P}$	cholesteric pitch
P_2	orientation function, defined in this text as $\mathscr{S}$
P_j	probability that first sites required by each of the remaining sequences be vacant in the Flory model
$P_n(\cos\alpha)$	spherical harmonic of order n
q	persistence length
R	gas constant
$\mathscr{R}$	rate of bond fracture
R_g	radius of gyration
S	entropy
S	order parameter
S^*	critical order parameter
s	disclination strength
s_c	statistical weight of coil sequences
s_z	scattering vector for diffraction $= 2\sin\theta_B/\lambda$
T^*	characteristic temperature
$\hat{T}$	temperature at which the persistence length is reduced to the cut-off length
T_g	glass transition temperature
$T_{lc\rightarrow i}$	liquid crystal to isotropic transition temperature
T_m	melting temperature
T_n	compensation temperature at which a cholesteric becomes nematic
t_i	distance from centre of gravity of chain to link i

U	potential energy per mole
u	potential energy
V_c	threshold voltage for alignment
V_{rms}	root mean square average value of AC voltage
v	velocity
v_p	volume fraction of polymer
v_p'	concentration of isotropic phase in biphasic region
v_p'	critical volume concentration of polymer for onset of liquid crystallinity
v_p''	concentration of anisotropic phase in biphasic region
v_p^*	polymer volume fraction at which a minimum in the partition function first appears
v_s	volume fraction of solvent
W_i	weight fraction of molecules with molecular weight M_i
W_s	anchoring energy
X_i	number fraction of molecules with molecular weight M_i
x	axial ratio
x_c	cut-off axial ratio
x_K	persistence ratio of a Kuhn chain
x_p	persistence ratio
y	misorientation of a rod in the Flory model
y	number of sequences in the Flory model
Z	partition function
Z_{comb}	combinatory part of the partition function
Z_{orient}	orientational part of the partition function
Z_{en}	'energetic' part of the partition function
Z_{rot}	rotational part of the partition function
Z_{conf}	conformational part of the partition function
z	coordination number
α	angle between molecule and director
α	polarizability
α_i	ith Leslie coefficient of viscosity
β	second order molecular susceptibility
ε	dielectric tensor
$\varepsilon_\perp$	relative permittivity along molecular axis
$\varepsilon_\parallel$	relative permittivity perpendicular to molecular axis
η	viscosity
η^*	viscosity at concentration v_p''
$\eta_{a,b,c}$	Miesowicz coefficients
η_s	solvent viscosity
$\eta_{T,S,B}$	viscosities associated with twist, splay and bend respectively

Θ	equilibrium fraction of helical units
θ	angle rod-like molecule makes with the director in the Flory model
θ_B	Bragg angle
λ	wavelength
λ_0	reflective wavelength
μ	total dipole of molecule
μ_i	molecular susceptibility
v	frequency
v_j	number of positions at which the jth rod can be added to the lattice in the Flory model
ρ	number density of chains
$\rho(\alpha)$	orientation probability function
σ	statistical weight of helical sequences
$\sigma, \boldsymbol{\sigma}$	stress tensor
τ	relaxation time
$\tau, \boldsymbol{\tau}$	shear stress tensor
χ	Flory–Huggins interaction parameter
χ_a	diamagnetic anisotropy
$\chi_\perp$	magnetic susceptibility along molecular axis
$\chi_\parallel$	magnetic susceptibility perpendicular to molecular axis
Ω	optical rotatory power
ω	angle director makes with layer normal in smectic C
ω_y	solid angle associated with misorientation y in the Flory model
ΔG	change in free energy
$\Delta\chi$	anisotropy of magnetic susceptibility
Δn	birefringence
$\Delta\omega$	change in bonding energy on dissolution
*	chiral centre
$\langle r^2 \rangle$	mean square end-to-end distance

Molecule Index

<1> CH_3-O—⬡—$\overset{O}{C}$—O—⬡—$\overset{O}{C}$—O—⬡—NO_2

<2> CH_3-O—⬡—$N\overset{\uparrow O}{=}N$—⬡—$O-CH_3$ PAA

<3> CH_3-O—⬡—$CH=N$—⬡—$(CH_2)_3-CH_3$ MBBA

<4>
$$R_4-\overset{R_1}{\underset{R_3}{\overset{|*}{C}}}-R_2 \qquad R_4-\overset{R_1}{\underset{R_2}{\overset{|*}{C}}}-R_3$$

<5> CN—⬡—⬡—⬡—$(CH_2)_m-\overset{H}{\underset{CH_3}{\overset{|}{C^*}}}-C_2H_5$

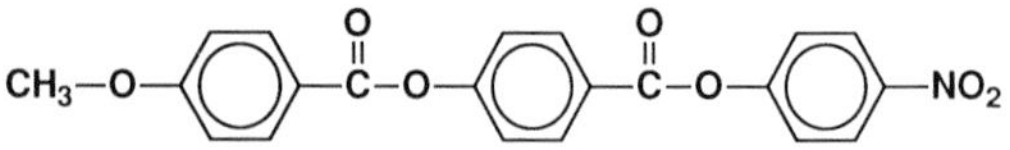

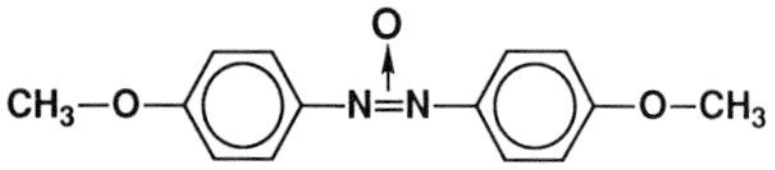

<6> $CH_3-(CH_2)_{12}-\overset{O}{C}-O$—(steroid)

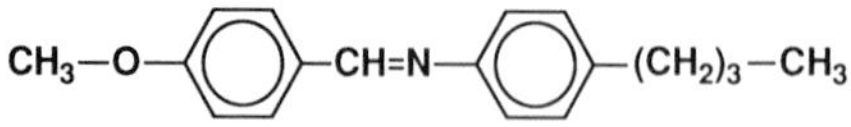

<7> $C_6H_{13}-O$—⬡—$CH=N$—⬡—$O-C_6H_{13}$

<8(i)>
$$\begin{array}{l} CH_2=CH \\ \overset{}{O}=C \\ \quad O-(CH_2)_6-O \end{array}$$—⬡—⬡—CN

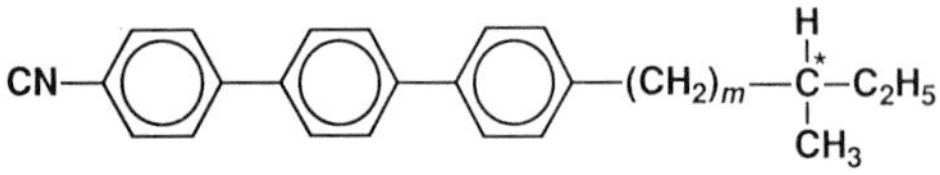

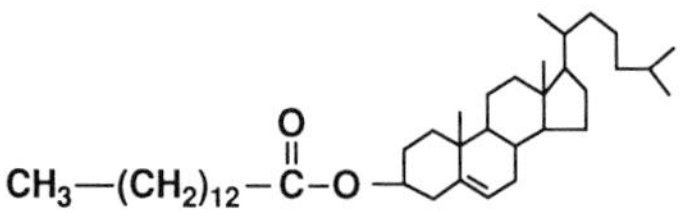

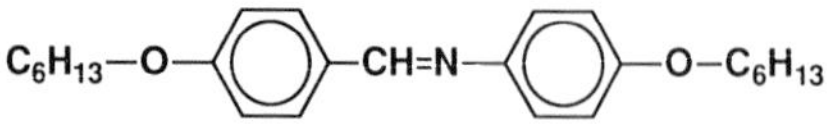

<8(ii)>
$$\left[\begin{array}{l} -CH_2-CH- \\ O=C \end{array} \right]_n$$
$$O-(CH_2)_6-O$$—⬡—⬡—CN

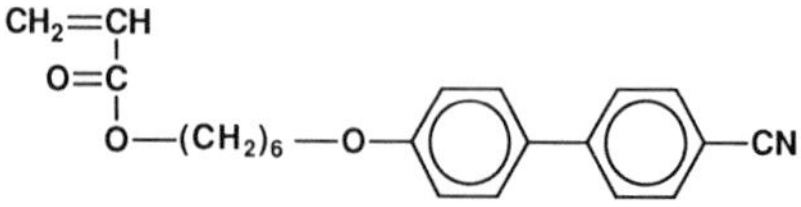

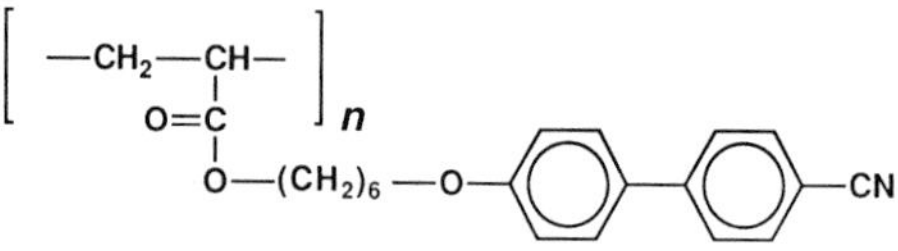

<9>

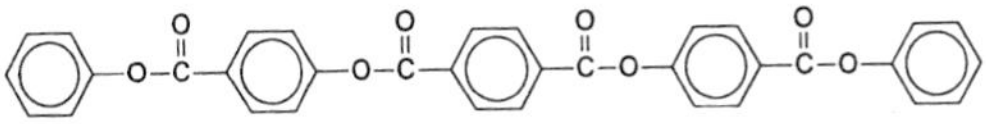

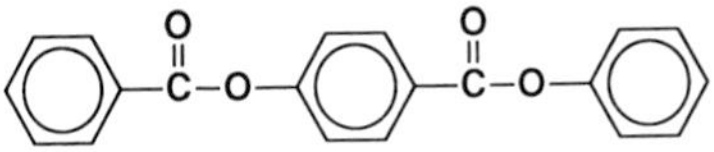

<10>

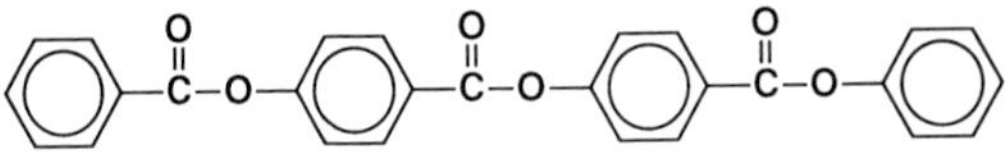

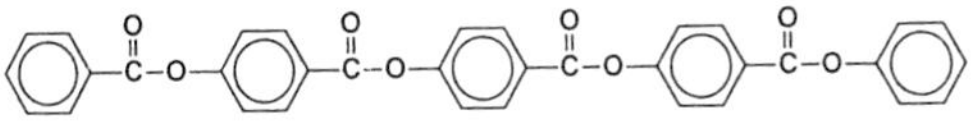

<11>

<12>

<13>   PHBA
poly(hydroxybenzoic acid)

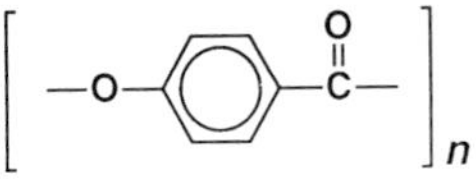

<14>   PHNA
poly(hydroxynaphthoic acid)

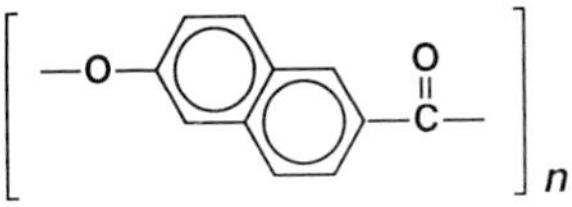

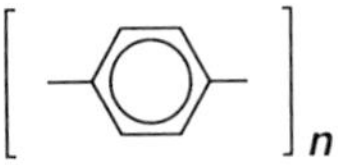

<15>   PPP
poly(*p* - phenylene)

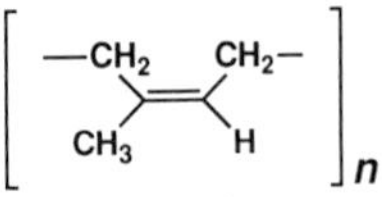

<16>   polyisoprene

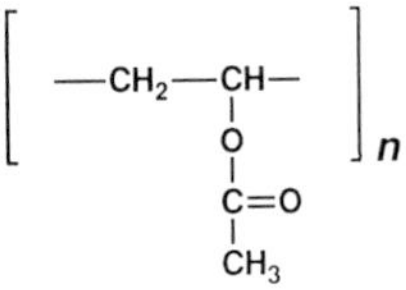

<17>   poly(vinylacetate)

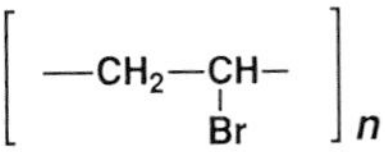

<18>   poly(vinylbromide)

<19>   PTFE
poly(tetrafluorethylene)

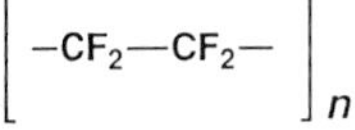

<20> R is NO_2 — cellulose nitrate

<21> R is $-CH_2CHCH_3$ (OH) — HPC hydroxypropyl cellulose

<22> R is $-(CH_2)_2-C(=O)-O-CH_2-$ (phenyl) — PBLG poly(γ-benzyl-L-glutamate)

<23> PBA poly(p-benzamide)

<24> PPT poly(p-phenyleneterephthalate)

<25> PET poly(ethyleneterephthalate)

<26>

<27>

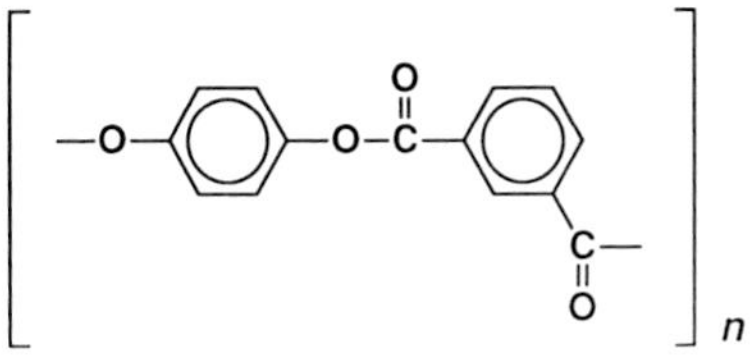

<28>

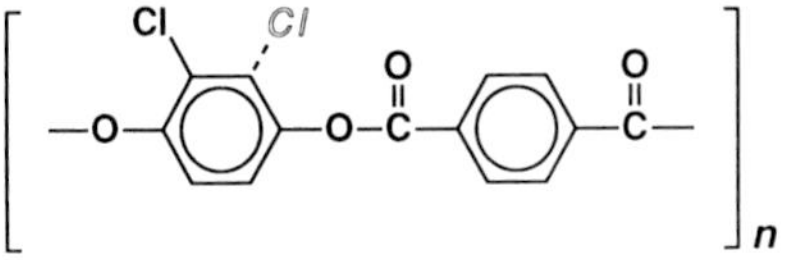

<29>

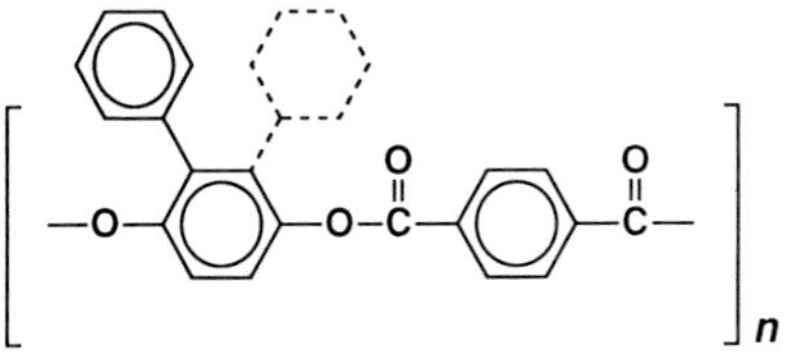

<30>

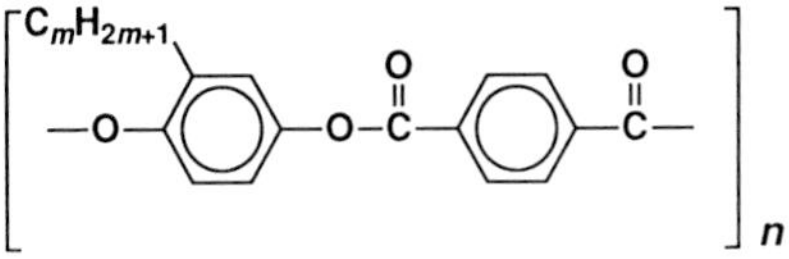

<31>

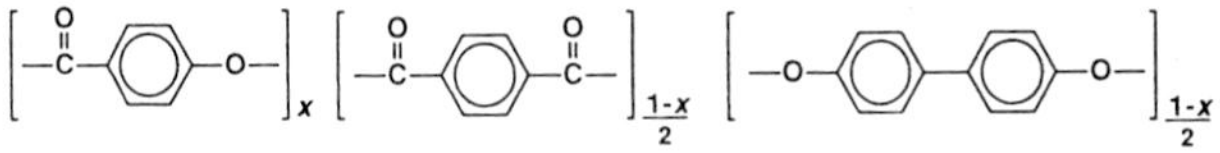

<32>

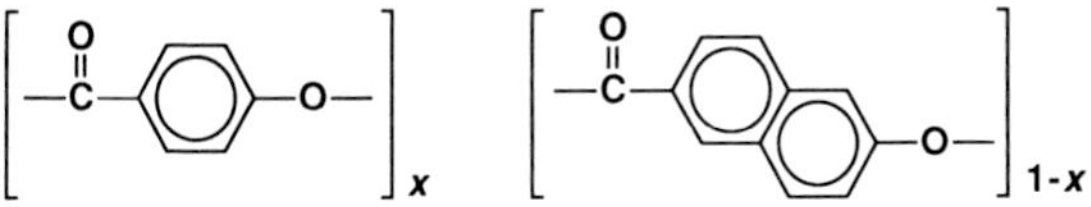

<33>

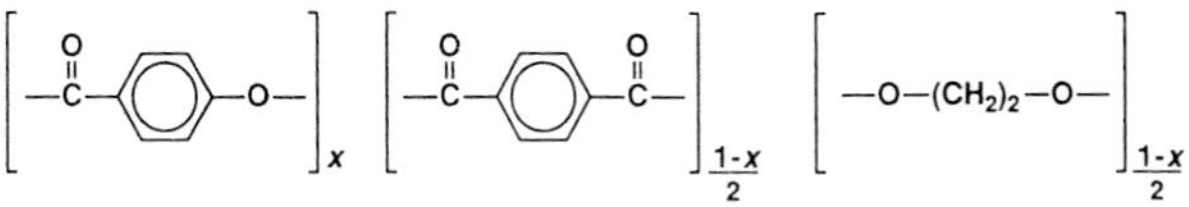

<34> $\left[\begin{array}{c} \text{structure} \end{array}\right]_x$ $\left[\begin{array}{c} \text{structure} \end{array}\right]_{\frac{1-x}{2}}$ $\left[\begin{array}{c} \text{structure} \end{array}\right]_{\frac{1-x}{2}}$

<35> $\left[\begin{array}{c} \text{structure} \end{array}\right]_x$ $\left[\begin{array}{c} \text{structure} \end{array}\right]_{\frac{1-x}{2}}$ $\left[\begin{array}{c} \text{structure} \end{array}\right]_{\frac{1-x}{2}}$

<36> $\left[\begin{array}{c} \text{structure} \end{array}\right]_{0.25}$ $\left[\begin{array}{c} \text{structure} \end{array}\right]_{0.25}$ $\left[\begin{array}{c} \text{structure} \end{array}\right]_{0.5}$

<37> $\left[\begin{array}{c} \text{structure} \end{array}\right]_{0.5}$ $\left[\begin{array}{c} \text{structure} \end{array}\right]_{0.25}$ $\left[\begin{array}{c} \text{structure} \end{array}\right]_{0.25}$

<38> $\left[\begin{array}{c} \text{structure} \end{array}\right]_{0.5}$ $\left[\begin{array}{c} \text{structure} \end{array}\right]_{0.25}$ $\left[\begin{array}{c} \text{structure} \end{array}\right]_{0.25}$

<39> $\left[\begin{array}{c} \text{structure} \end{array}\right]_{0.5}$ $\left[\begin{array}{c} \text{structure} \end{array}\right]_{0.25}$ $\left[\begin{array}{c} \text{structure} \end{array}\right]_{0.25}$

<40> $\left[\begin{array}{c} \text{structure} \end{array}\right]_{0.5}$ $\left[\begin{array}{c} \text{structure} \end{array}\right]_{0.25}$ $\left[\begin{array}{c} \text{structure} \end{array}\right]_{0.25}$

<41> $\left[\begin{array}{c} \text{structure} \end{array}\right]_{0.5}$ $\left[\begin{array}{c} \text{structure} \end{array}\right]_{0.25}$ $\left[\begin{array}{c} \text{structure} \end{array}\right]_{0.25}$

<42> $\left[\begin{array}{c} \text{structure} \end{array}\right]_n$

<43> C₂H₅—O—⟨benzene⟩—C(CH₃)=CH—⟨benzene⟩—O—C₂H₅

$C_2H_5-O-\phi-\underset{CH_3}{C}=CH-\phi-O-C_2H_5$

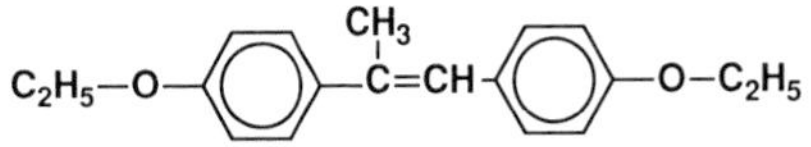

<44>

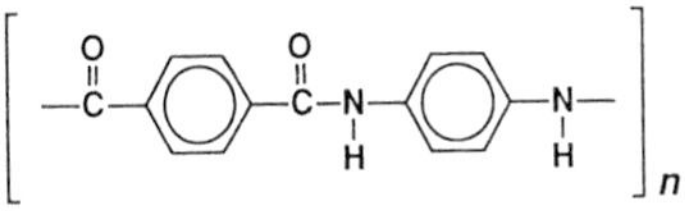

PPTA
poly(*p*-phenylene
terephthalamide)

<45>

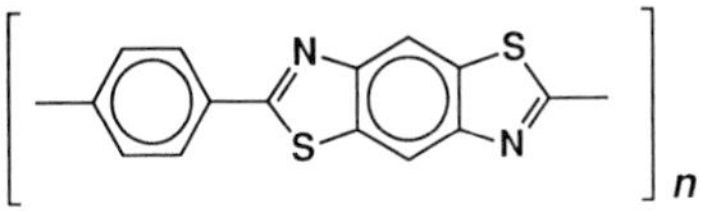

PBTZ
poly(*p*-phenylene
benzobisthiazole)

<46>

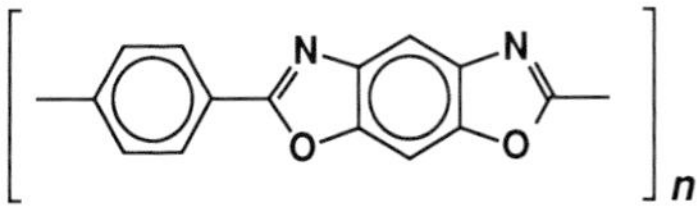

PBO
poly(*p*-phenylene
benzobisoxazole)

<47>

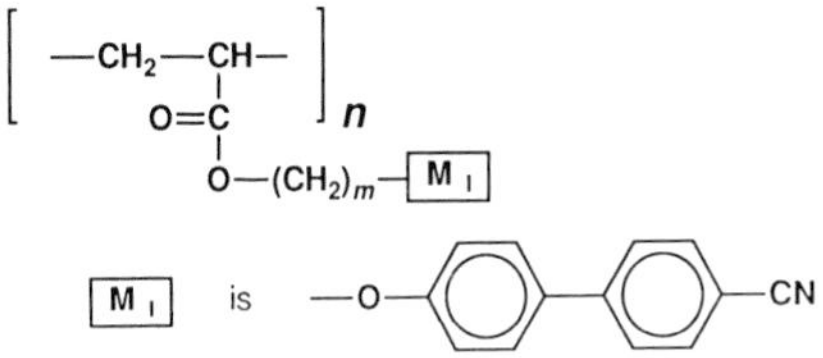

M_I is $-O-\phi-\phi-CN$

<48>

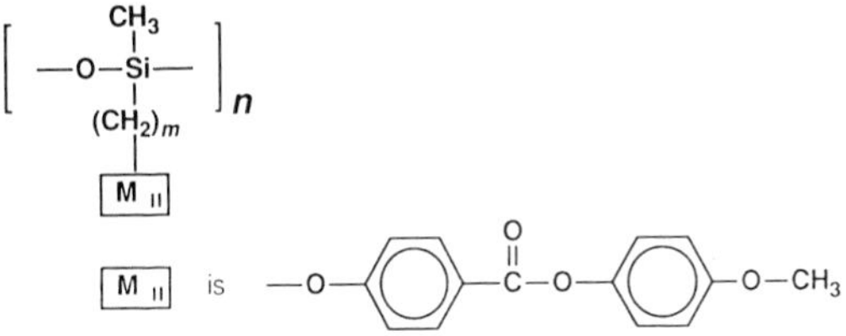

M_{II} is $-O-\phi-\overset{O}{C}-O-\phi-O-CH_3$

<49>

<50> $CH_3-O-\phi-CH=N-\phi-N=CH-\phi-O-CH_3$

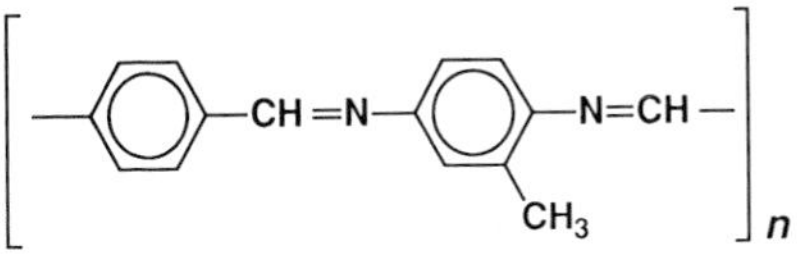

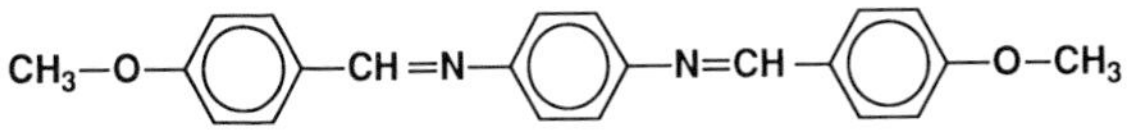

<51>

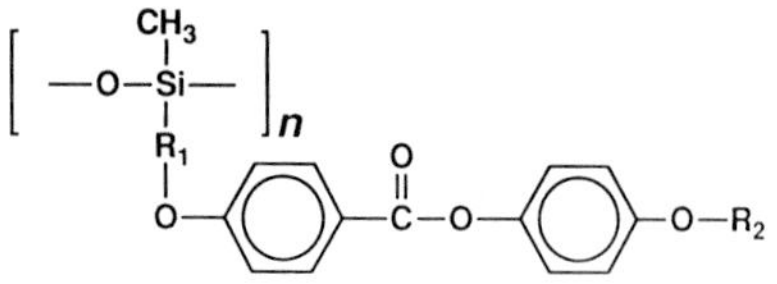

$$\left[\begin{array}{c} CH_3 \\ -O-Si- \\ R_1 \end{array} \right]_n$$

<52>

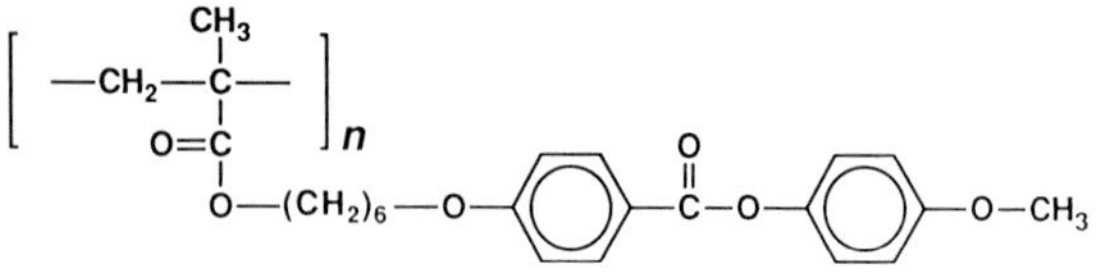

$$\left[\begin{array}{c} CH_3 \\ -CH_2-C- \\ O=C \end{array} \right]_n$$

<53>

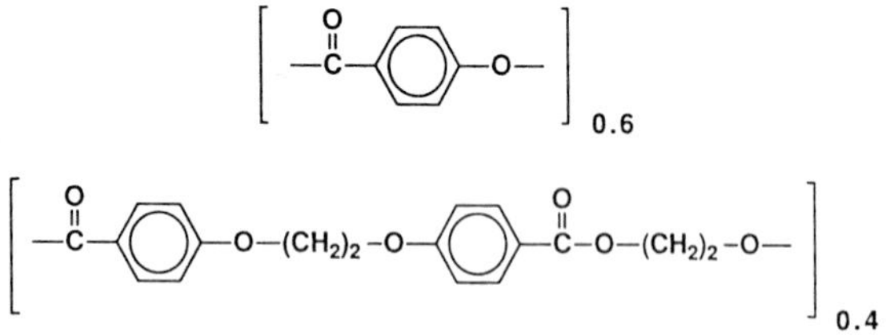

$$\left[\begin{array}{c} O \\ -C-\bigcirc-O- \end{array} \right]_{0.6}$$

$$\left[\begin{array}{c} O \\ -C-\bigcirc-O-(CH_2)_2-O-\bigcirc-C-O-(CH_2)_2-O- \end{array} \right]_{0.4}$$

<54>

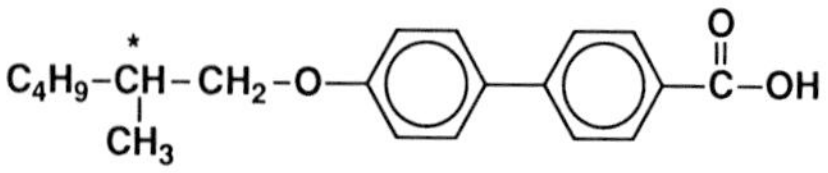

$$C_4H_9-\overset{*}{C}H-CH_2-O-\bigcirc-\bigcirc-\overset{O}{C}-OH$$
$$\quad\quad\quad CH_3$$

*   Chiral centre

<55>

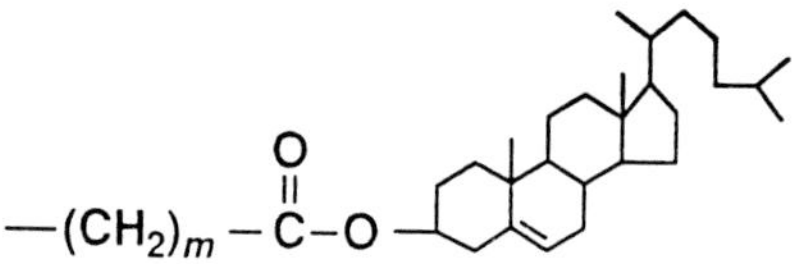

$$-(CH_2)_m-\overset{O}{C}-O-$$

<56>

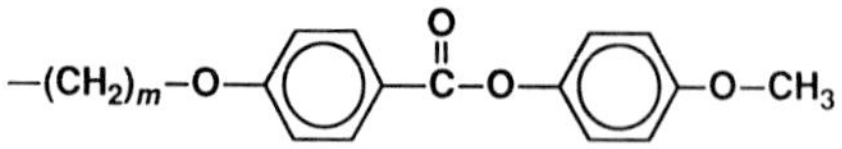

$$-(CH_2)_m-O-\bigcirc-\overset{O}{C}-O-\bigcirc-O-CH_3$$

<57>

$$\left[\begin{array}{c} CH_3 \\ -CH_2-C- \\ O=C \end{array} \right]_n$$

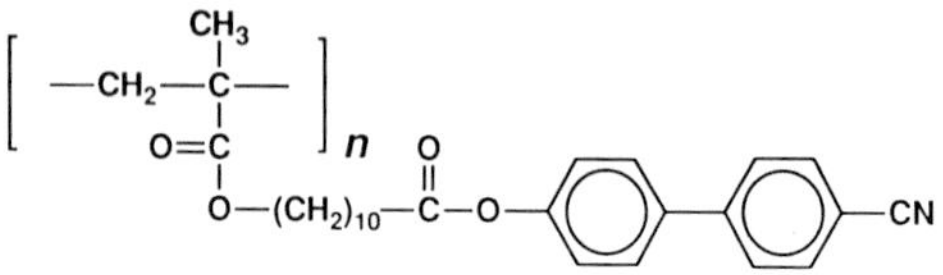

<58>

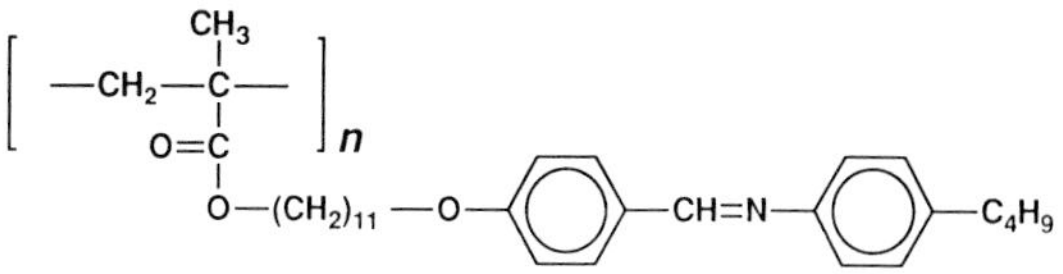

$$\left[-CH_2-\underset{\underset{O=C}{\overset{CH_3}{|}}}{C}- \right]_n$$

$O-(CH_2)_{11}-O-\langle\text{benzene}\rangle-CH=N-\langle\text{benzene}\rangle-C_4H_9$

<59>

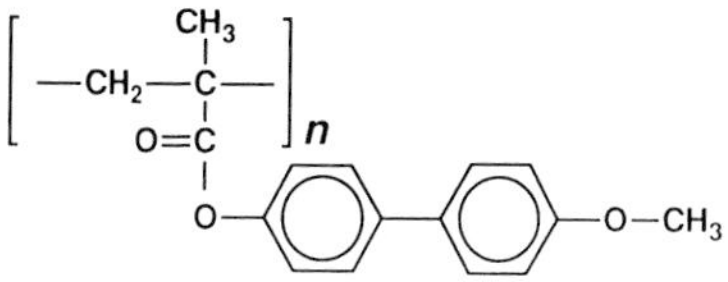

$$\left[-CH_2-\underset{\underset{O=C}{\overset{CH_3}{|}}}{C}- \right]_n$$

$O-\langle\text{biphenyl}\rangle-O-CH_3$

<60>

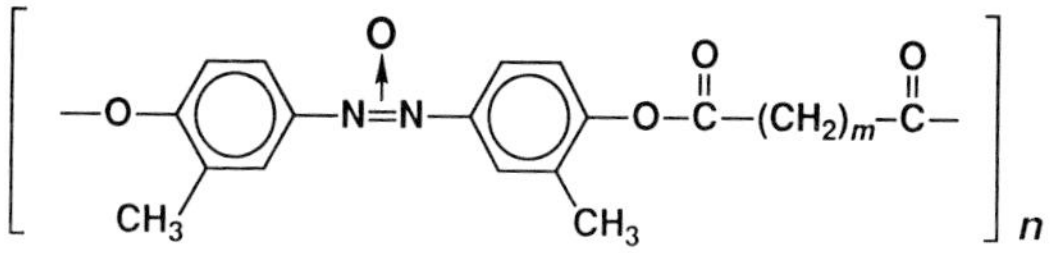

$$\left[-O-\langle\text{benzene, }CH_3\rangle-\underset{O}{\overset{\uparrow}{N}}=N-\langle\text{benzene, }CH_3\rangle-O-\underset{O}{\overset{\parallel}{C}}-(CH_2)_m-\underset{O}{\overset{\parallel}{C}}- \right]_n$$

<61>

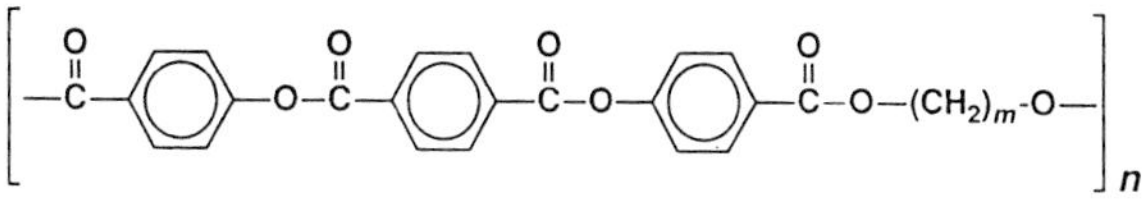

$$\left[-\underset{O}{\overset{\parallel}{C}}-\langle\text{benzene}\rangle-O-\underset{O}{\overset{\parallel}{C}}-\langle\text{benzene}\rangle-\underset{O}{\overset{\parallel}{C}}-O-\langle\text{benzene}\rangle-\underset{O}{\overset{\parallel}{C}}-O-(CH_2)_m-O- \right]_n$$

<62>

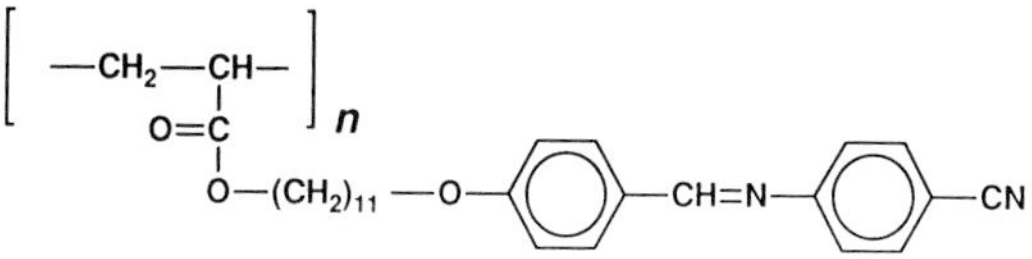

$$\left[-CH_2-\underset{\underset{O=C}{}}{CH}- \right]_n$$

$O-(CH_2)_{11}-O-\langle\text{benzene}\rangle-CH=N-\langle\text{benzene}\rangle-CN$

<63>

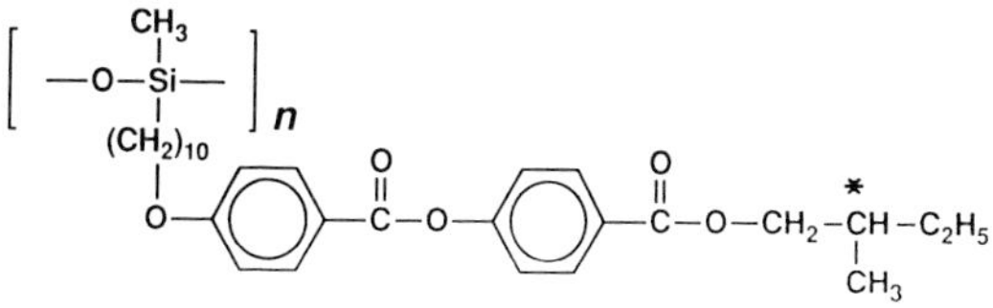

$$\left[-O-\underset{\underset{(CH_2)_{10}}{\overset{CH_3}{|}}}{Si}- \right]_n$$

$O-\langle\text{benzene}\rangle-\underset{O}{\overset{\parallel}{C}}-O-\langle\text{benzene}\rangle-\underset{O}{\overset{\parallel}{C}}-O-CH_2-\overset{*}{C}H-C_2H_5$
$\quad\quad\quad\quad\quad\quad\quad\quad\quad\quad\quad\quad\quad\quad\quad\quad | $
$\quad\quad\quad\quad\quad\quad\quad\quad\quad\quad\quad\quad\quad\quad\quad CH_3$

<64>

$$\left[-O-\langle\text{benzene, }Cl\rangle-O- \right]_{0.25} \left[-\underset{O}{\overset{\parallel}{C}}-\langle\text{benzene}\rangle-\underset{O}{\overset{\parallel}{C}}- \right]_{0.25} \left[-O-\langle\text{benzene}\rangle-O- \right]_{0.25}$$

$$\left[-\underset{O}{\overset{\parallel}{C}}-\langle\text{benzene}\rangle-O-(CH_2)_2-O-\langle\text{benzene}\rangle-\underset{O}{\overset{\parallel}{C}}- \right]_{0.25}$$

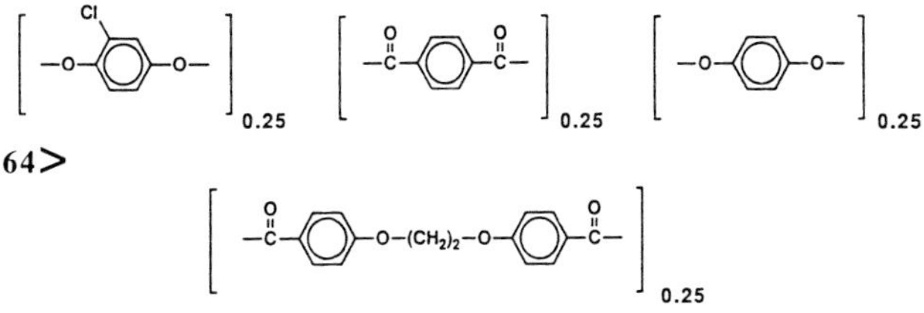

<65>

$$-(CH_2)_n-O-\!\!\langle\bigcirc\rangle\!\!-\overset{\overset{\displaystyle O}{\|}}{C}-O-\!\!\langle\bigcirc\rangle\!\!-R$$

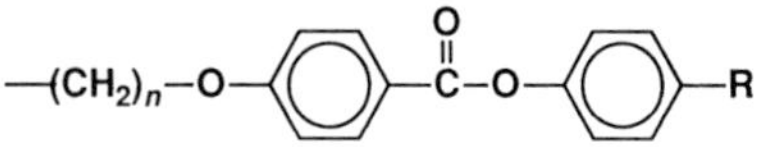

<66>

$$\left[-(CH_2)_m-\overset{\overset{\displaystyle O}{\|}}{C}-O-\!\!\langle\bigcirc\rangle\!\!-\overset{\overset{\displaystyle O}{\|}}{C}-O-\!\!\langle\bigcirc\rangle\!\!-O-\overset{\overset{\displaystyle O}{\|}}{C}- \right]_n$$

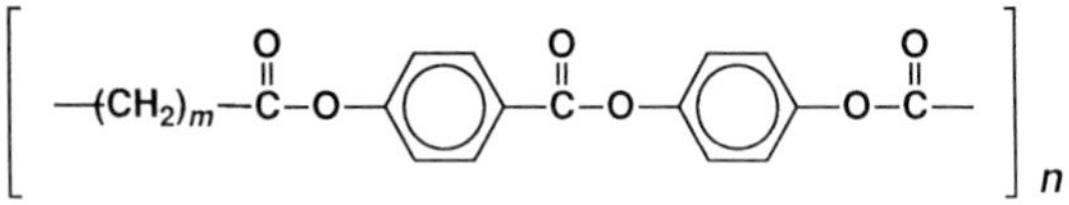

<67>

<68>

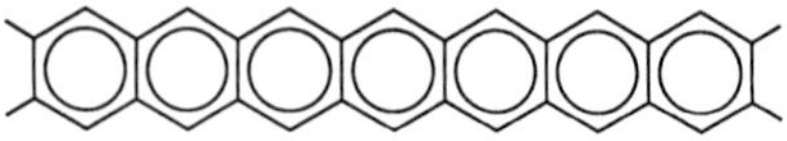

$$\left[-C\equiv C-\overset{\overset{\displaystyle R}{|}}{C}=\overset{\underset{\displaystyle R}{|}}{C}- \right]_n$$

<69>

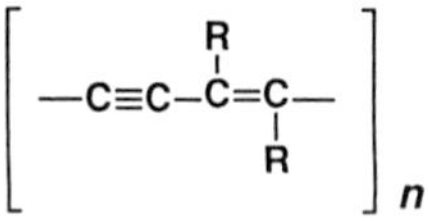

$$NH_2-\!\!\langle\bigcirc\rangle\!\!-NO_2$$

<70>

$$N(CH_3)_2-\!\!\langle\bigcirc\rangle\!\!-\overset{\overset{\displaystyle CN}{|}}{C}=\overset{\overset{\displaystyle}{}}{\underset{\underset{\displaystyle CN}{|}}{C}}\!\!-CN$$

<71>

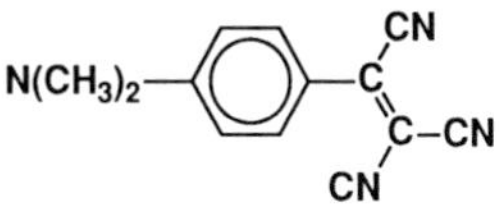

$$\overset{\overset{\displaystyle CN}{|}}{C}=\underset{\underset{\displaystyle CN}{|}}{C}-CN$$

<72>

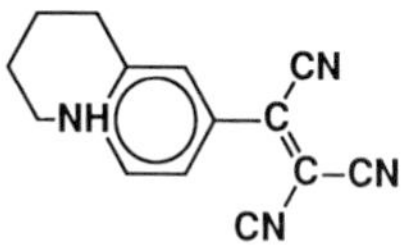

$$N(CH_3)_2-\!\!\langle\bigcirc\rangle\!\!-CH=N-\!\!\langle\bigcirc\rangle\!\!-NO_2$$

<73>

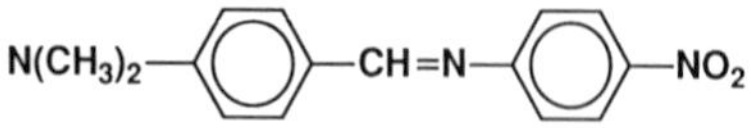

$$N(CH_3)_2-\!\!\langle\bigcirc\rangle\!\!-CH=CH-\!\!\langle\bigcirc\rangle\!\!-NO_2$$

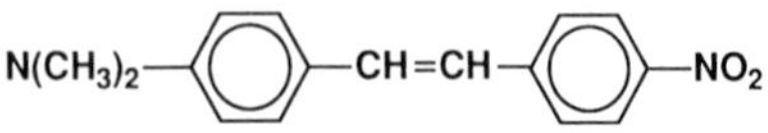

<74> NH_2—⟨⟩—⟨⟩—NO_2

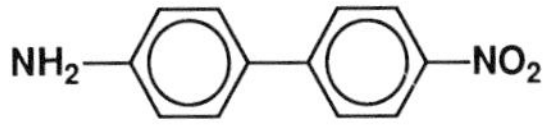

<75> NH_2—⟨⟩—⟨⟩—⟨⟩—NO_2

<76> $N(CH_3)_2$—⟨⟩—$CH=CH-CH=CH$—⟨⟩—NO_2

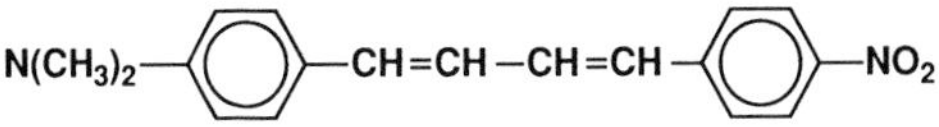

<77>
$$CH_2=\overset{\displaystyle CH_3}{\underset{\displaystyle \underset{O}{\overset{\parallel}{C}}}{C}}$$
O—$(CH_2)_{12}$—O—⟨⟩—⟨⟩—NO_2

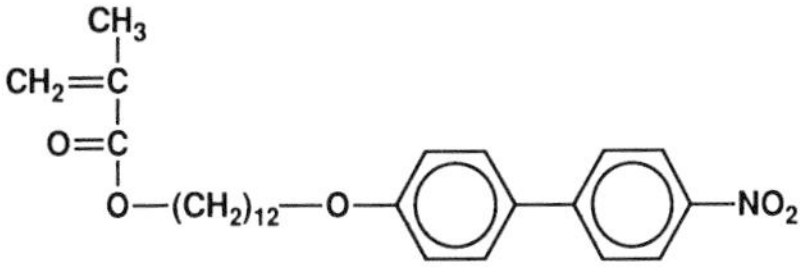

<78>
$$\left[-CH_2-\overset{CH_3}{\underset{\underset{O}{\overset{\parallel}{C}}}{C}}- \right]_x$$
O—$(CH_2)_n$ —O—⟨⟩—⟨⟩—NO_2

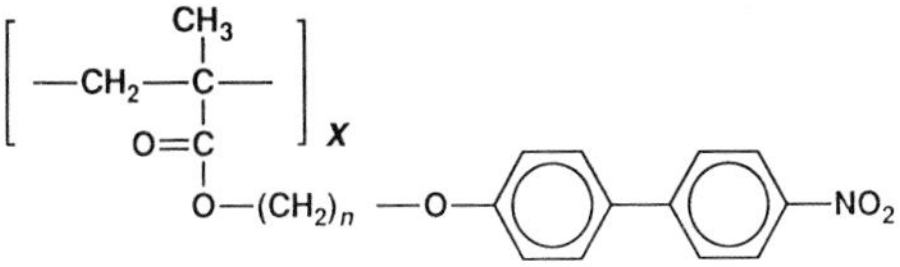

<79>
$$\left[-O-\overset{CH_3}{\underset{(CH_2)_n}{Si}}- \right]_x$$
O—⟨⟩—⟨⟩—NO_2

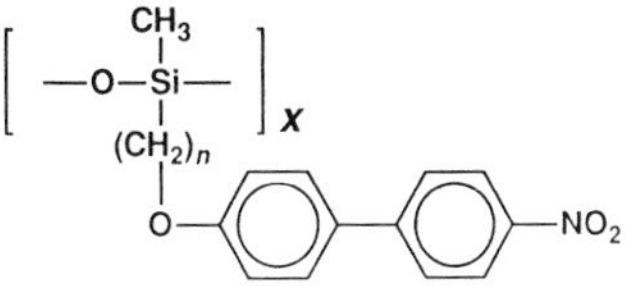

<80>
$$\left[-CH_2-\overset{CH_3}{\underset{\underset{O}{\overset{\parallel}{C}}}{C}}- \right]_n$$
O—$(CH_2)_{10}$—$\overset{O}{\overset{\parallel}{C}}$—$O$—⟨⟩—$O$—$\overset{O}{\overset{\parallel}{C}}$—⟨⟩—$\overset{O}{\overset{\parallel}{C}}$—$O$—$CH_2$—$\overset{*}{CH}$—$CH_2$ $\overset{CH_3}{\underset{CH_3}{}}$
$\underset{Cl}{}$

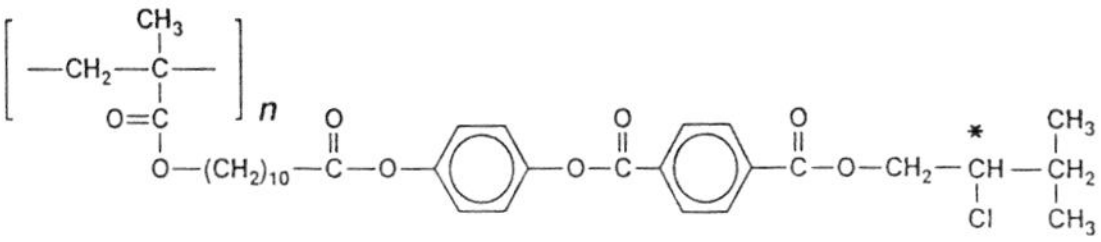

<81>
$$CH_2=CH$$
$$\overset{\parallel}{\underset{O}{C}}$$
O—$(CH_2)_2$—O—⟨⟩—$\overset{O}{\overset{\parallel}{C}}$—$O$—⟨⟩—$CN$

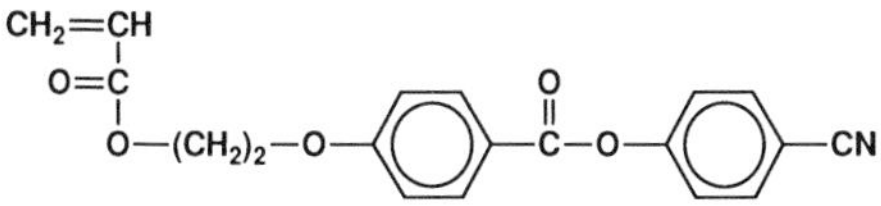

<82>
$CH_2=CH$ $CH=CH_2$
$\overset{\parallel}{\underset{O}{C}}$ $\overset{\parallel}{\underset{O}{C}}$
O—$(CH_2)_2$—O

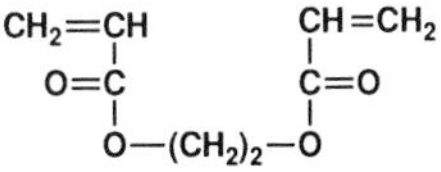

General index

amphiphillic 24
anchoring energy 185
aperiodic crystals 71, 176
atactic molecule 47
axial ratio 93, 115

banded textures 222f
bâtonnets 161, 194
bend 16, 171
bend constant (continuum) 118
bend constant (molecular) 242
biaxial figure 133
biaxial nematic 128
biphasic chimney 78
biphasic microstructures 196
biphasic region 77, 99ff
Bloch wall 188
block copolymer 8
bound solvent 65
broken focal conic 161
brushes 130, 180

coagulation bath 218
coefficient of thermal expansion (CTE) 247
chain dimensions 5
chirality 23, 25, 137
chiral smectic 25, 162
chiral smectic devices 278f
cholesteric 21, 23, 137, 189
circular dichroism 145
cis conformation 284
combinatory partition function 94
common tangent construction 45, 100
configuration 7
conformation 7
conformational energy 61
conjugation 266, 270
conoscopic image 132
contour length 5
copolymer 7
Cotton–Mouton effect 264
crankshaft rotation 8, 61, 248
creep 252
critical concentration 98ff, 203, 212
critical field 174, 177
crystal solvate 81
cybotactic nematic 155

DSC (differential scanning calorimetry) 9
decoupling 87
dernier 31
degree of polymerization 33
dielectric spectroscopy 227, 249
die-swell 220
diffraction 11
dipolar alignment 20, 268f
director 16, 38
disclination 178, 182, 186, 189
dislocation 178, 189
displacement (vector) 20, 35
Doi–Edwards model 209f
drawdown 219, 257
Dupin cyclide 192f

elastic constants 18, 171, 200
elastic modulus 31, 236
elastomeric liquid crystals 265, 280f
electric field 19, 225f
electroactivity 264ff
elongational flow 222
entanglements 210, 239
enthalpy 44
entropy 45, 237, 262
equator 13
Ericksen–Leslie theory 204ff
escape in the third dimension 183
extended conformation 238
extrapolation length 185
extrusion 220

fan textures 160, 196
ferroelectric behaviour 278
fibre spinning 218ff
fibrillation 232, 254
fingerprint textures 142, 174
first normal stress differences 203
flexible spacers 51, 73, 225
flexural modulus (of mouldings) 260
Flory–Huggins parameter 46, 78
focal conics 142, 160, 191f, 282
fountain effect 220
fractionation 106
Frank constants 18, 171
Fredericks' deformations (transition) 176, 200, 208f